Decision support systems in potato production

Decision support systems in potato production

bringing models to practice

edited by:
D.K.L. MacKerron
A.J. Haverkort

Wageningen Academic
P u b l i s h e r s

CIP-data Koninklijke Bibliotheek
Den Haag

ISBN 9076998302

Subject headings:
Chain management
Information technology
Agriculture

First published, 2004

Wageningen Academic Publishers

The Netherlands, 2004

Printed in The Netherlands

Table of contents

Preface

Decision support systems in potato production: bringing models to practice

In March 2003 the third international potato modelling conference took place in Dundee. The first such conference took place thirteen years ago in Wageningen. Its subject then was to compare potential and water limited yields calculated by several existing simulation models of temperature driven development and radiation and water driven growth. Developments in modelling were so fast that within five years at the second conference researchers presented models and their yield prediction of potato crops subjected to many biotic and abiotic factors. They included late blight, wilting disease, nematodes and limitations due to reduced availability of water, nitrogen, and adaptation ranges of temperature, radiation and daylength. Then and since, quantitative crop ecological information has been used both to enhance the scientific models and to test their robustness over various potato ecologies.

With increasing interest of modellers in applying the models in the field and of both consultants and growers in getting the benefits from models, more and more models are now being developed into the bases of quantitative decision support systems (DSS). The development is fuelled by increased confidence that model calculations approach reality, provided that appropriate input data are available. The theme of the third international potato modelling conference was easily found, therefore. The participants of the conference were all active in developing and testing DSS and were asked to submit a chapter to this book. The first, introductory chapter 1 shows what kind of decisions are supported and the future role of DSS in potato supply chain management. The other chapters each give a description of a DSS, the science on which it is based, illustrative examples of its computer screen displays and, where applicable, some users' responses. Each concludes with a summary table allowing appraisal in a quick glance. Chapter 2 on zoning shows where and when potatoes can be grown at what yields levels to assist companies and governments. The chapters 3 - 5 deal with nitrogen DSS from various angles, requiring destructive and non-destructive crop sampling, whereas chapter 6 shows a DSS using only modelling to assist growers to quantify and time their applications of water and nitrogen. A major quality aspect - tuber size distribution - is treated in chapter 7. The management advisory package for potato, MAPP, which is already in commercial operation is treated extensively in chapter 8 and its working is illustrated with seed rates, yield prediction, irrigation scheduling and tuber grade distribution. DSS on crop protection is shown in chapter 9 on rotational aspects of potato cyst nematodes, chapter 10 similarly for all nematodes and their geo-referenced distribution, chapter 11 on chemical control of late and early blight, and chapter 12 on chemical weed control. The book concludes with a chapter, 13, on designing ideal genotypes for certain conditions as a DSS for breeders and a final chapter, 14, that encourages the scientist to focus on practicalities, to include only necessary and sufficient variables, and to produce a balance between accuracy and realism. It draws on a few examples presented in the earlier chapters.

The editors

1. Role of decision support systems in potato production

A.J. Haverkort and D.K.L. MacKerron

Among the many potato research establishments throughout the world there exists a wealth of expertise in a wide range of aspects of potato culture from agronomy to pathology. Public sector research establishments need to disseminate their scientific findings to the sectors that they serve and so facilitate wealth creation. Some commercial companies have similar aims. Unfortunately, that expertise is largely distributed (unfocused) and rarely, if ever, has it been presented in a coordinated or coherent form. Meanwhile, potato growers have been seeking ways to combine the results of research and development in a form that they can use and certain growers complain that they have no way of accessing the extensive knowledge that exists among experts. Staff at some research establishments have developed models, implemented in computer programmes that simulate several aspects of potato growth and that address some of the problems faced by growers. But, again unfortunately, almost all of these computer programmes are quite unsuited to the non-specialist. Developments in modern information technology and the general acceptance of computers with standard operating systems mean that much of that knowledge and many of these computer programmes can be combined into user-friendly decision support systems (DSS). The *knowledge is available* and now the *technology is also available* to present that knowledge. The time is ripe for the development of integrated DSS that will combine usefully whole sectors of knowledge into single advisory packages mounted in computer-based software, that would give potato growers access to the best knowledge available on culture of potato and that would assist with their management decisions. This book presents accounts of several advances made recently in this direction.

Types of decisions

Strategic decisions

The most strategic decision taken in potato production (Figure 1) is to grow them at all. The desirability and probability of acceptable yields can be studied in agro-ecological zoning DSS. Once the decision to grow crops has been made, ideotyping is a strategic DSS that may help one decide which variety to grow. Of course, in traditional potato-growing areas, experience accounts for the first of these and market demand may limit the second.

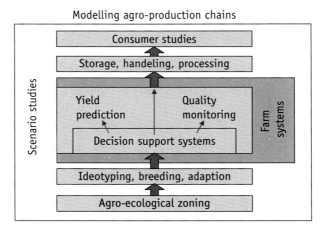

Figure 1. Examples of the role of modelling and decision support systems.

Synthetically produced nitrogen fertilizer only became available in the second half of the twentieth century. This made it possible for farmers to match the demand and supply of the crop more precisely than before. Previously, organic manure was the main source of nitrogen for the potato crop and in some areas it has remained important. The management of soil organic matter, generally to maintain or increase its levels, is crucial for a number of reasons. Soil organic matter improves the physical condition of the soil, increasing aggregate stability, improving the water holding capacity, and conserving the soil. It also improves the chemical soil fertility as it can provide a long-term, if unsteady supply of plant nutrients. The best strategic management and planning to make optimal use of soil organic matter depends on the type of farm (arable, mixed, integrated, organic) its size and the possibilities to vary the rotation. DSS on soil fertility and fertilization assist in making the strategic decisions regarding organic matter amendments. As well as influencing the quantity and quality of organic matter, crop rotation influences the dynamics of soil borne organisms such as weeds, nematodes and diseases. DSS exist to forecast behaviour of such organisms depending on soil type, crop management and varieties used.

Tactical decisions

Tactical decisions are ones made in the in the year potatoes are grown and call for information from that year. For instance what a farmer needs to know before actually applying nitrogen to a particular field in a specific year are: how much nitrogen the soil will supply, how much the crop will demand. Soil analysis at the end of winter partly informs the first question but in addition one needs to know the expected rate of mineralization of nitrogen from soil organic matter, which depends on the type of organic matter previously incorporated in the soil, and any losses (leaching) dependent upon the pattern of rainfall. The demand by the crop will depend on its eventual yield but the potential yield also depends on the weather and the availability of water, possibly from irrigation. A late maturing crop has a longer growing season, generally has a higher yield and, so, has a higher nitrogen demand than early or short-season crops. Early and short-season crops may be grown from early maturing cultivars or may be grown from mid-season cultivars killed prematurely, as is the case with seed potatoes or when the market calls for 'new potatoes'. The sum effect of all these considerations is to make it difficult, if not impossible to make effective, crop-specific decisions on nitrogen application without a computer-based decision support system.

Further one must decide whether to apply all the nitrogen fertilizer at planting or to apply it in split doses. Tactical decisions on when and how to apply nutrients and crop protection agents are critical in avoiding losses of nitrogen and emission of active ingredients to the environment. Tactical DSS include how to manage the water supply to the crop (whether to irrigate and what kind of irrigation system) and when and how to sow catch crops to make use of the nitrogen in the next growing season.

Operational decision support

Once the crop has been planted, the choice of variety and seed rate has been made, and any organic manure has been applied followed by an initial dressing of nitrogen and other minerals, the operational decisions remaining to be taken by the farmer are those on supplemental nitrogen, water supply, crop protection, and when to stop or kill the crop for harvest. Support for these decisions can come from an array of monitoring and sampling techniques for the soil and the crop. Figure 2 shows the complexity of the decision making process. For nitrogen DSS, invasive methods include the destructive sampling of plant parts to determine their concentrations of nitrate and total nitrogen. Recently, non-invasive techniques have been developed from which measurements correlate well with nitrogen concentration and, even, uptake. Such techniques help farmers to determine whether or not the crop needs additional fertilization. Soil sampling helps to determine whether the soil nitrate pool is nearing depletion. Supplemental dressings should be given taking these measurements into account. The dressings are only useful when the additional nitrogen reaches the plant in time to extend its growth. Foliar dressings are helpful, being rapidly, but incompletely absorbed. Soil dressings require rainfall or irrigation shortly after application if they are to be available to the plant. They are taken up less quickly and, again, will not all be absorbed by the crop. Experience is gained with the combined application of water and nitrogen through fertigation in several countries. As figure 2 shows, most DSS have been developed for the field production stage of the potato supply chain. The major aim of these DSS is to assist growers in their operational decisions.

Use of DSS to increase sustainability in potato chains

There is no single supply chain for potatoes, rather there are many supply chains. Figure 3 schematically represents such a supply chain. The consumers are on the top of the

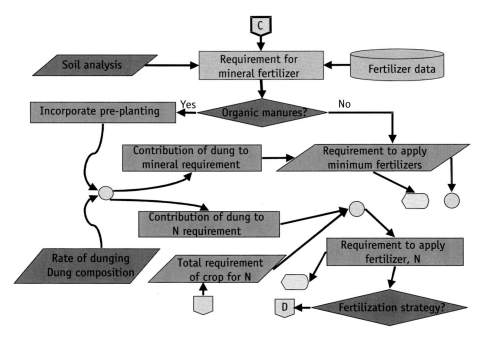

Figure 2. The increased complexity of quantitative decision making (Karvonen et al., 2000).

chain and the breeders of new varieties form the foundation of the chain. A supply chain is situated in a physical and socio-economic environment from which it draws resources and to which it supplies material. The strongest flow of information is from the consumers as a driving force down the chain. Information is also going the other way through tracing and tracking to inform the consumer of the quality of the produce and the impact it may have had on the environment. The material flow is from breeding to consumers as potato seed, seed potatoes and processing potatoes.

Agricultural production chains increasingly aim at ensuring food and product safety. The CIES Global Food Safety Initiative is an example thereof and shows the pre-competitive nature. Tracing and tracking, the establishment of certificates (EUREPGAP), hazard analysis and legal accountability (EGFL) are known endeavours. Ensuring a constant high quality of food and products at a convenient time, enhancing economic growth and international competitiveness is the second important aim of the industry. Integrated Pest Management (IPM) is considered a food safety item and not competitive. Ensuring sustainability is next in importance but more competitive as striving for it may affect product quality and efficacy of the production chain. Potato production is sustainable if the three P's requirements are met. If one of the three requirements is not met, the industry loses its "licence to produce". People should benefit from healthy nourishment and socially acceptable production methods, there must be Profit in it for all links of the chain and the Planet should not suffer.

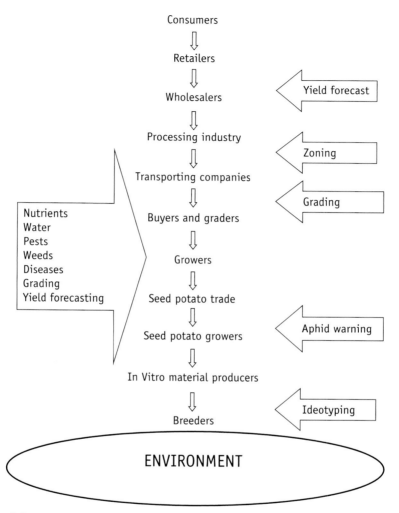

Figure 3. All links of a well-developed potato supply chain within its environment; horizontal arrows show examples of existing decision support systems.

Integrated management of pests, crops and farms

DSS-based integrated pest management (IPM) is an endeavour that is mainly driven by consumer-demand aimed at reducing the use of pesticides, fungicides and herbicides to reduce the risk of residues in the consumed produce. Examples are: the detection and distribution of nematodes and subsequent planting of nematode-resistant varieties or local application of nematicides; systems for the detection and control of epidemics of aphids and of late blight; systems for application of minimum effective doses of herbicide. Integrated crop management (ICM) reaches farther and also optimises the use of water and nutrients with the aid of irrigation planners and nutrient application decision support systems based on yield expectation and soil or crop mineral contents. Integrated farm management reaches still farther and takes into account crop rotation to assure soil fertility (organic matter management) and soil health. These management practices increasingly make use of quantitative approaches such as models and databases, techniques for remote sensing, measurement and detection, and information and communication technology. Management data increasingly are reinforced with crop data required by the industry for tracing and tracking purposes, by the authorities for environmental purposes and generic data from e.g. meteorological services. Farm management systems are being developed at present to relate the data with scientific insights into self learning systems.

Increasingly it is observed that labels and certificates require that, if a DSS exists for certain management decisions that may have an impact on the environment, they are then used! This movement is likely to become the main driving force behind the development of DSS

DSS in supply chain management

The links of potato supply chains are not totally commercially and financially interdependent. Breeders may breed varieties for a specific company through a contract but more often they are free to choose the possibility to use breeders rights as a source of income. The same applies for seed production and the part of a ware potato crop that is not contracted. Managing more and more links in a potato chain is becoming more and more important. A well managed chain reduces flows of waste products and uses fewer services and fewer providers, leading to cost reduction and a competitive edge. Food safety scandals in the past have made consumers and producers aware of the necessity to assure food safety at all levels of production: breeders must make sure that they breed varieties with low levels of toxic glycoalkaloids, producers must make sure no residues of toxic chemicals occur in the harvested produce and that tuber nitrate concentration is low. The use of resources should be such in potato production, storage, transport and processing that resources are used as efficiently as possible. Here DSS play a crucial role.

The various players (links) within the potato chain are interdependent and yet may operate independently if they desire and can afford it. Mechanisms to enforce communication to steer the chain in the desired direction are the following:
- supply and demand through market movements (including stock market options)
- enforced quality assurance (e.g. seed quality SE, E, A, B) and label requirements (EUREPGAP, Assured Produce,..) increasingly require DSS to be applied when available
- processing company requirements regarding food safety, sustainability indicators and quality (e.g. dry matter, defects, size, shape,...) are steered with DSS
- quality control when purchasing by each link (e.g. seed health certification and contractual control to assure desired quality levels for processing)

- tracing and tracking: loads of potatoes and their products purchased by consumers ideally should be traced back several steps up the chain. Much required quantitative information is stored in DSS.

Potato chain management is rapidly gaining importance for more efficient, safer, more sustainable production. This process is strongly aided by the revolution in Information and Communication Technology (ICT) and progress made in sensing technologies (crop scanning) and in decision support systems (DSS). Figure 4 schematically shows how consumers steer processing industries and how their factories increasingly plan their raw material procurement using DSS. Factory raw material requirements indicate which field should grow what variety, its timing and rate of inputs and moment of harvest or storage. It is still partly futuristic, but the developments of DSS, ICT and up-scaling of farms and the industries are such that automated processes throughout the chain are becoming reality.

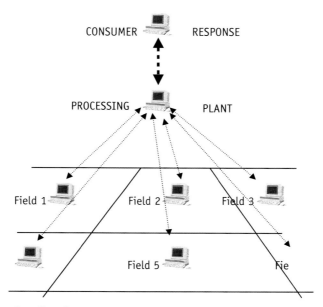

Figure 4. Information flow from consumer to field. Each computer represents a quantitative decision support system.

Future developments in decision support

Sensing and models

There is an increased need for decision support systems for a number of reasons. With application of water, fertilizers, and agro-chemicals being subject to restriction, it is important to make the fullest use of available knowledge in order to make the best use of less or fewer resources, additionally, DSS can be used as evidence that applications of an agro-chemical were justified. A second reason is that as farms become larger the farmers benefit more from assistance with their decisions. These decisions are valid for increasingly larger areas and so cost less per unit area. Thirdly, and maybe most importantly, improved access to knowledge as such and to knowledge aggregated in models, and the availability of sensing techniques and of information and communication technology can improve the services given by extension agents. Lastly, governments are rapidly reducing their direct support of the primary production sector of society, especially the farming community. Many private consultancy firms are being formed and they will serve farmers best if tools are available for decision support. These four reasons are likely to become more important in the coming years.

Models are abstractions of reality, reductions, and yet a lot of quantitative information is needed to run them. Most of it is generic information, not unique to that crop, such as weather and soil parameters. Rate variables in any model that is likely to be used in decision support systems include the conversion rate of solar energy into dry matter and the water use efficiency. Both of these depend on the availability of water, solar radiation, nitrogen and biotic stresses. Many of the quantitative interactions between these variables are known or are rapidly being included in current models. For decision support, models need more inputs than just generic data. They need crop-specific data such as on planting and emergence and the progress simulated by a model should be confirmed by observation to give confidence in the predictions of water or nitrogen uptake and of yield. This is where sensing techniques come into their own; techniques such as infrared reflectometry for proportional ground cover, electronic probes for soil water content, and ratios of reflection of various wavelengths of light to determine the nitrogen concentration of the standing biomass. Calculation of response to a certain amount of input about to be applied will require that a model that previously ran on current data should switch to some estimate of likely future weather such as long term averages. Such a procedure would also allow the calculation of risks by calculating consequences of treatments for years that are drier or wetter than average.

Information and communication technology

Increasing amounts of data are collected in the course of crop production. Some of these are site- and time-specific and are produced as a consequence of decisions such as nitrogen and water requirements within the course of a growing season. Traceability in production chains also requires an increased amount of data to be registered. Similarly, the requirement to meet certain standards for product registration and certification, as in organic farming, demands additional record keeping.

There is, now, an increased desire to capitalise on these data and to turn them into a self learning system. Nowadays many of the data are stored in digital form. Extension firms serving farmers with decision support systems keep most information on disease prevention and levels of nitrogen and water in soil and crops available in automated computer-based systems. Records of yield and quality (dry matter and nitrogen concentration, frying colour, protein content,..) are similarly kept by the processing

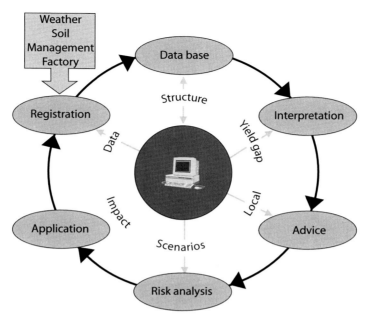

Figure 5. Schematic representation of a self learning or self adjusting system making use of automated generation and transfer of knowledge (Haverkort and MacKerron 2000).

industries. Such field-specific information is ideally suited to a self learning system that gets better after each year in which data are collected (Figure 5) as the decisions may be tailored to an individual site. Figure 5 shows the requirements and working of such a system. Data are collected annually on a field and stored in a database. With the aid of quantitative approaches (statistical and mechanistic models) the data are interpreted using yield-gap analysis and other techniques and suggestions are generated for time, site and dose of inputs. Before being implemented, this advice is tested in scenarios with long-term average weather data for robustness and risk of failure (too many or too few sprays, over or under irrigation and over or under nitrogen fertilization). Once the recommendation has been implemented the response is recorded for subsequent years and the database, and subsequent recommendations, are adjusted accordingly.

An advantage of the whole system is that a continuum is created between farmers, extension agencies collecting and interpreting the data, and science because all three parties exchange data and views and learn and improve while contributing to the system.

References

Haverkort, A.J. & D.K.L. MacKerron, 2000. Recommendations and trends in research and practice of application of nitrogen and water to potato crops. In A.J. Haverkort & D.K.L. MacKerron (Eds) Management of nitrogen and water in potato production. Wageningen Pers: 301-309.

Karvonen, T., D.K.L. MacKerron & J. Kleemola, 2000. Role of simulation and other modelling approaches in decision making. In A.J. Haverkort & D.K.L. MacKerron (Eds) Management of nitrogen and water in potato production. Wageningen Pers: 250-262.

Name of the DSS

POTATO ZONING

Who owns the DSS, Principal author's, address,...

Anton Haverkort

Plant research International, Wageningen UR

P.O. Box 16 6700 AA Wageningen

anton.haverkort@wur.nl

What questions does it address?

- where and when can you grow potatoes?
- what kind of yields are expected?
- what quality (grades, dry matter concentration) is to be expected?
- what are the risks (drought, heat, frost)?

What input is needed?

Long term weather data such as minimum and maximum temperatures, rainfall, evapotranspiration, soil water holding capacity.

Who is it for, who are the intended users?

This decision support system is intended for:

- national governments and research institutions to set agendas for research and development
- companies interested in producing seed potatoes
- companies wanting to build a processing factory in a new area.

What are the advantages over conventional methods (whatever those may be)

Conventional methods include trying to grow (seed)potatoes in an area and find out over time whether growers adopt the technology and whether the crop expands or shrinks. This method using POTATO ZONING leads to assess yields and risks of hazards to crops using long term weather data . The more accurate these data are the better yield, quality and risks can be pre-determined.

How frequently should the DSS be used or its values be updated?

Once all relevant weather and soil data have been acquired the DSS calculates expected long term yields and variation between past years. After each potential growing season the DSS can use past season's data and make a new run

What are the major limitations?

The model assumes that for each environment (site x climate) where yields are calculated a suitable variety exists (somewhere). Such a variety may exist yielding the calculated amounts but such a variety not necessarily may need the demands of the users.

What scientific / technical enhancements would be desirable?

To estimate more accurate yield levels the light use efficiency may need to be made more dependable on light intensity. Also so far the model uses mean day/night temperatures. The performance of the model likely ill benefit from using daily minimum and maximum temperatures. Reduced grid size will increase its usefulness for most users. Finally, more extensive groundthruthing needs to be carried out by comparing actual yields achieved with model calculation at many different sites.

2. Potato-zoning: a decision support system on expanding the potato industry through agro-ecological zoning using the LINTUL simulation approach

A.J. Haverkort, A. Verhagen, C. Grashoff and P.W.J. Uithol

POTATO-ZONING detects and describes agro-ecosystems suitable for potato. It is of use for the rapidly expanding potato processing industry and for potato research and development projects. It answers questions such as: Where can potatoes be grown? What is the expected length of the growing season? What yields are expected with and without irrigation? What are the risks of drought and night frosts? What is the quality (dry matter concentration and tuber size distribution) going to be? And what happens to yields if potatoes are bred with increased tolerance of night frost or heat?

POTATO-ZONING is based on the crop growth model LINTUL (Light Interception and Utilization of Light) that calculates dry matter production and yields based on data on temperature, solar radiation and water availability. Using a global weather set containing 30-year average weather in a 30 x 30 arc minutes pixel grid, POTATO-ZONING calculates potential and water limited yields for whole continents and allows a quick scan of where irrigation may increase yields and whether potato producing areas are near populated areas. Zooming in on regions and using a soil data set with a 5 x 5 arc minutes pixel grid POTATO-ZONING calculates regional trends in quality aspects and the benefits of increased tolerance of climatic hazards.

Introduction

The potato industry rapidly is becoming a global industry with links between countries that exchange through import and export of seed potatoes, ware and processed potatoes and genetic material. Figure 1 shows where potatoes are grown at present. There are strong increases evident in Asia and Africa and there is some reduction in eastern and central Europe. In particular, the processing industry, producing crisps and deep frozen chips, expanded rapidly in the western hemisphere in the last decades of the twentieth century. Currently the greatest expansion in the production of these products is ocurring in Asia. When moving into a new area the processing industry will ask itself the two main questions:

- Is there a market?
- Can we grow potatoes?

Subsequent questions are:
- How large is the potato production area?
- What kind of varieties can be grown?
- What yields can be expected with and without irrigation?
- Can we vary the growing season within the year
- Can we store the tubers after harvest under ambient conditions?
- What quality can we achieve regarding e.g. dry matter concentration, tuber size?
- Can seed potatoes be grown locally or do they have to be imported?
- How remote are the growing and processing areas from the consumers? (The market and production should be close as to avoid expensive transport.

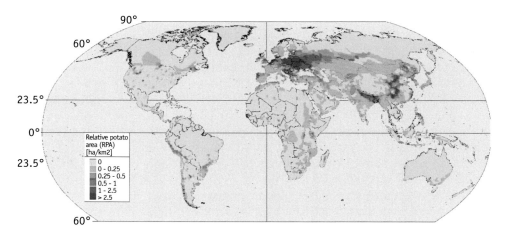

Figure 1. Global potato distribution (Hijmans, 2001).

Figure 2 illustrates what happened in the United States of America. Over one hundred years ago potatoes were grown where people lived. Now potatoes are produced where they can best be grown given the climate and soil conditions. So a survey of where potatoes are actually grown (Figure 1) is not necessarily the best indicator of the optimal growing area. When expanding the potato industry in existing or to new areas in the world the questions above need to be answered. The decision support system on expanding the potato industry through agro-ecological zoning using LINTUL illustrated in this paper is then a helpful tool.

Scientific basis

Potato crop development such as sprout growth rate, emergence, leaf area development depend on temperature. Dry matter accumulation depends on the amount of solar radiation intercepted by the crop and dry matter distribution between the various organs is determined by temperature and daylength. When there is less water available than

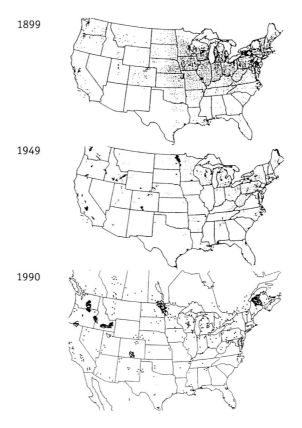

Figure 2. Distribution of potato production in 1899, 1949 and 1990 (Hawkins, 1957; Rowe, 1993)

needed for optimal growth, growth is reduced. Van Keulen and Stol (1995) used the potato crop growth model LINTUL (Light Interception and Utilization of Light, Spitters 1987) to calculate yields at any given place and season. POTATO-ZONING uses the same assumptions consisting of:

- A potato crop grows only between temperatures of 5 and 21°C. With daily average temperatures below 5 °C there is too much risk of frost and development is too slow.
- With daily average temperatures (day and night) above 21°C it is too hot for potato production. It is assumed that in many growing areas this coincides with day-night temperatures of 15-27 °C. At such temperatures daily growth is reduced due to increased respiration, dry matter distribution mainly favours the foliage and tuber dry matter concentration above 21°C becomes unacceptably low.
- Based on the above temperature window and with a base temperature of 2°C, the thermal time or temperature sum (units, Kelvin days, K d) of a growing season is calculated in an area. If during a season with temperatures above 5°C and below 21

°C less than 1250 Kelvin days are accumulated the season is considered too short and no yield will be calculated. If on the other hand the thermal time exceeds 2200 K d the calculations assume there to be two distinct growing seasons.

- The model is initiated by defining the plant density and the initial light interception by the foliage at emergence and the initial relative growth rate of that foliage. The fraction of intercepted (F_{int}) photosynthetically active radiation (PAR between 400 and 700 nm wavelength) is calculated from subsequent foliar increase. The model assumes a 2-week period between planting at suitable temperatures and emergence of the crop.
- Senescence of a crop leads to a linear reduction of light interception from 1 to zero over a period of 600 K d from the end of the season.
- The length of the growth cycle of the varieties are expected to match exactly that of the available length of the growing season, resulting in a harvest index of 0.80 at crop senescence but is reduced at higher temperatures to reach 0 at 28 °C.
- Daily growth (W) is calculated as W = LUE * F_{int} PAR, with LUE = 2.9 g MJ^{-1} (Haverkort, 1990). In the previous study (Stol *et al.*, 1991), Light Use Efficiency (LUE) was not dependent on temperature. However, POTATO-ZONING accounts for this based on present ecophysiological knowledge (Kooman & Haverkort, 1995). The base value of LUE is multiplied by a reduction factor TF, its value of depends on the daily mean temperature as shown in Figure 3.
- Roots are expected to grow 60 cm deep and precipitation (infiltration), percolation, soil surface evaporation and crop transpiration are calculated depending on soil type and crop development. POTATO-ZONING starts running on the day of planting with a soil water content of 80% of field capacity. This is based on the assumption that, after winter or after a dry period, farmers will plant as soon as the field is neither

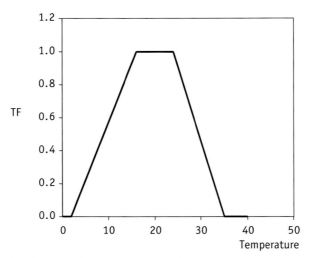

Figure 3. Reduction factor TF (see text) versus daily mean temperature. This factor accounts for the effect of temperature on LUE (light use efficiency) (Kooman & Haverkort, 1995).

too dry nor too wet. Obviously the model will make mistakes for areas where it never rains such as in deserts where 80% of field capacity will not be reached.
- At an LAI (leaf area index) of 4, transpiration is maximal. Where the actual/potential transpiration ratio is smaller than 1 the crop accumulates drought stress linearly affecting LUE and when its cumulative sum exceeds 10 it linearly decreases light interception assuming that leaves are shed.
- For *regional studies,* POTATO-ZONING uses two simple relations between environment and potato quality derived from Haverkort & Harris (1987). The dry matter concentration of tubers diminishes by 0.446 percentage points per °C and the number of tubers per plant increases by 1.68 tuber, each per °C increase in temperature. Base values were defined as 20% dry matter and 12 tubers, at 14 °C. These relations were linked to the relation between temperature and elevation derived from the regional data set.

POTATO-ZONING defines agro-ecosystems for potato on a global scale using data sets on soil (FAO, 1996) and climate (Leemans & Cramer, 1991; Müller, 1996). It establishes the spatial distribution of the length of the growing season, potential production and water-limited production. Global climatic data sets used are compared to a selection of local stations, and the consequences for the selection of temperature thresholds is briefly discussed. A world digital elevation map is used to compare with the calculated spatial patterns.

Approach

POTATO-ZONING as presented here is a follow-up of the potato production and zoning study performed in 1991 by Stol *et al.* Their simulation model is now linked to a geographical information system (GIS) environment. Updated global data sets and regional data sets are used which allow a more detailed analysis.

For both the *global studies* and the *regional studies*, POTATO-ZONING uses long-term weather data from two sources. Leemans & Cramer (1991) published a 30 x 30 arc minutes pixel grid on a global scale with long-term average monthly data of cloudiness, temperature and precipitation. The degree of cloudiness is used to calculate solar radiation. As the model uses daily data as input, the monthly averages need to be interpolated. For radiation and temperature this is a straightforward procedure. For the distribution of precipitation an intermediate step is needed in the statistical distribution of the monthly total rainfall over the total number of days that it rains. These rainy days, however, are not mentioned in the databases of Leemans & Cramer (1991). Therefore they are derived from the number of rainy days of 1272 meteorological stations compiled by Müller (1996 and subsequently they are translated to the 30 x 30' grid.

Global soil data set

The water availability used in *global studies* is based on the FAO soil map published on CD-ROM (FAO, 1996). This map classifies the soil types of the world based on a pixel grid of 5 x 5 arc minutes. POTATO-ZONING uses three texture classes to determine the water holding capacity of the soil. Even within the small 5 x 5 arc minute pixel several soil types may be present. Therefore, water-limited yield is calculated using weather data from the 30 x 30 grid as a weighted average based on the relative proportion of each soil type per 5 x 5 pixel. As the FAO soil map only distinguishes three major soil texture classes, POTATO-ZONING takes the data from these three classes from Stol *et al.* 1991 (Table 1).

Table 1. Volumetric water content at field capacity and wilting point, and water holding capacity (cm m^{-1}).

Water characteristic	Soil type		
	Coarse	Medium	Fine
Field capacity	13	32	54
Permanent wilting point	4	10	44
Water holding capacity	9	22	10

Regional data set

For more detailed *regional studies*, POTATO-ZONING uses additional information from the METEO database of Plant Research International of Wageningen University and Research Centre, the Netherlands. This contains two sources: first, the long-term Climatic Data from the FAOCLIM Global Weather Database (FAO, 1990), second, the Handbuch ausgewählter Klimastationen der Erde (Müller, 1987). The POTATO-ZONING approach in the *regional study* uses weather station data for a precise zonation of current potato cultivars. For this approach, data from regional weather stations were taken as a reference. In this regional data set, information on daily minimum temperatures and maximum temperatures is available; this allows quantification of the relation between maximum temperatures and average temperature for each separate region. Next, this quantification is combined with the elevation data set and with data on quality in a calculation procedure.

Elevation data

For *regional studies* POTATO-ZONING also uses the digital elevation model (DEM) GTOPO30, resulting from a collaborative effort led by the staff at the U.S. Geological Survey's EROS Data Center in Sioux Falls, South Dakota. Elevations in GTOPO30 are regularly spaced at 30 arc seconds (approximately 1 kilometre). Figure 4 shows a strong correlation between elevation on the one hand and maximum, average and minimum temperatures on the other. A maximum temperature of 28°C (the upper limit for potato growth used in the *global study*) is linked with an average temperature of 23°C and a minimum temperature of 18°C, at an elevation of 650 m.

Stol *et al.* (1991) defined the temperature window suited for potato growth as the period in which the daily minimum temperature is higher than 5°C and the daily maximum temperature is lower than 28°C. In the present study, these thresholds could not be used, as the new global data set only contains daily mean temperature values and not the necessary daily minimum and maximum temperature values. For this reason, in the agro-ecological zone study, potato production was defined between a lower threshold of an average temperature of 5 °C and an upper threshold of an average temperature of 21°C. This selection was made via an iterative process using expert judgement of potato producing regions in the world. Next, in the ideotyping approach as part of the *regional study,* the same temperature window was used as a reference.

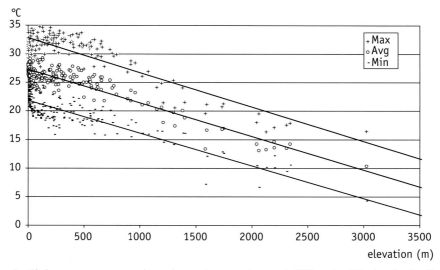

Figure 4. Minimum, average and maximum temperatures at different altitudes for India and South Asia.

Zoning explorations

Global explorations of potential and water limited yields

The results of calculations using POTATO-ZONING are presented on maps, each pixel representing an area of 0.5 x 0.5 arc degrees.

The calculations are shown here for the main potato season only. Where a second growing season is feasible, such as in the tropical highlands associated with the rainy seasons and in Mediterranean climates where a spring and an autumn crop can be grown, only the results of the main season are depicted in the maps. Yields are represented as tuber dry matter in tons per hectare (t dm ha^{-1}).

The Andean ecoregion, the genetic centre of origin of the potato (Figure 5) clearly has the highest potential yields of South America. Cloudless days with high incident solar radiation lead to high yields provided water is applied. Near the equator no potato production takes place at or near sea level due to inhibiting high temperatures in the Amazon basin. Yields increase with altitude in this area but the maps show that in the centre of the mountain range no potato production takes place due to too low temperatures above about 4000 metres above sea level year round. The further away from the equator the higher yields are due to long summer growing seasons with long days. The maps show that more remote from the equator increased areas of lower lying land are suitable. In the utmost south of the continent, potato production is not possible as the growing season is too short because of the long winters.

Figure 6 shows that the major potato growing areas in Africa are located near the Mediterranean Sea in Morocco, Algeria, Tunisia and Egypt where all crops are irrigated and only during part of the growing season receive some precipitation. This is the case for the early part of the spring season and the late part of the autumn season crop. Another important potato growing area is that associated with the mountain ranges of

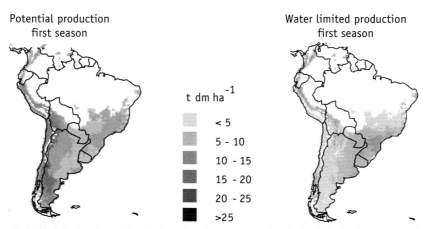

Potential production
first season

Water limited production
first season

t dm ha^{-1}

< 5
5 - 10
10 - 15
15 - 20
20 - 25
>25

Figure 5. Calculated potential (left) and water limited yields (right) for South America.

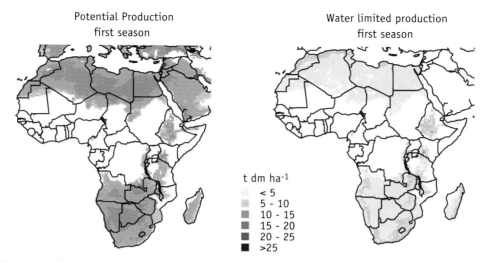

Figure 6. Calculated potential (left) and water limited yields (right) for Africa.

the Rift Valley stretching from the Bekaa Valley in Lebanon through Lake Malawi. There, most potatoes are grown under rainfed conditions in the highlands of Sudan, Ethiopia, Kenya, Uganda, Congo, Rwanda and Burundi. Rainfed yields there are the highest in the world and may exceed 20 t ha^{-1} tuber dry matter yield (over 80 t ha^{-1} fresh yield). Therefore, about 200,000 ha of potatoes are grown in those areas and potato has become an important staple food. A third large area is found in South Africa where conditions and cultural practices (irrigation) are not unlike those of Idaho, USA. Rainfed production is locally important in Madagascar and Cameroon and, similarly, irrigated small-scale production in the Sahel region Mali, Senegal and Cabo Verde.

The calculated yields for North America (Figure 7) much more closely resemble the actual situation in 1990 than that in 1899 (Figure 2) indicating that the potato is grown increasingly in areas where conditions for its production are most suitable. The southern areas of the United States, albeit having the highest potential yields have less potato

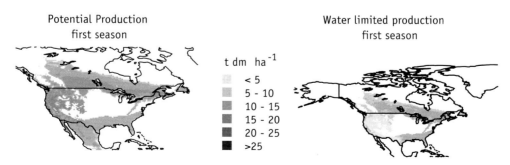

Figure 7. Calculated potential (left) and water limited yields (right) for North America.

than might be expected from figure 7 . Reasons may be the prices of land and irrigation water and high temperatures during the storage season.

Regional explorations of potato quality and tolerance

The regional data set is only used in the *regional studies*. From the available stations in the regional data set the relations of the yearly average of maximum temperature, average temperature and minimum temperature, versus the altitude were established (Figure 4) and subsequently used in POTATO-ZONING calculations.

Quality aspects

The most important quality aspects of potato are the tuber dry matter concentration and the tuber size. When tubers are initiated as swelling stolon tips the dry matter concentration is about 11 %. When tubers increase in size they also increasingly contain more starch and the dry matter concentration increases. The final dry matter concentration depends on cultivar, length of the growing season (the longer the higher the concentration), irrigation or rainfall (water availability decreases the dry matter concentration) and solar
intensity (low irradiation gives low yields and dry matter concentration). The average temperature during the growing season, however, is the main determinant factor for final dry matter concentration at harvest: low temperatures lead to high dry matter concentrations and vice versa. The main reason that potatoes are not grown in warm areas is that the dry matter concentration is too low - lower than 17 % is unacceptable

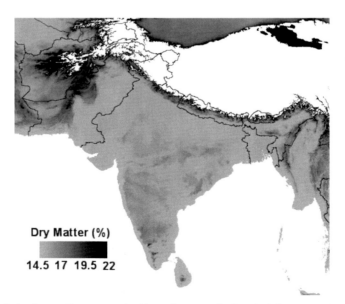

Dry Matter (%)

14.5 17 19.5 22

Figure 8. Trends in dry matter concentration of potato in South Asia.

due to poor storability and processing quality. Haverkort and Harris (1987) found a relation between dry matter concentration and altitude-dependent average temperature during the growing season: the dry matter concentration of tubers diminishes by 0.446% per °C temperature increase with a base value of 20 % at an average daily temperature of 14 °C. With this information, Figure 8 shows that the regions in South Asia capable of producing potatoes with acceptable dry matter concentration are concentrated in the Indo Gangetic Plain, Central India and South West India. Although North India has the potential for one growing season in e.g. West Bengal, Figure 7 shows that this region has a low potential for potato quality. Potatoes there are planted in November and harvested in February in which is a short, cloudy season with increasing temperatures in the second part of the tuber bulking period. Potential and actual yields are low and so are the dry matter concentrations.

Similarly, maps of the number of tubers per plant can be drawn with increasing temperature when growing potatoes at lower altitude. Haverkort and Harris (1987) found increasing number of tubers per plant: the number of tubers per plant increases by 1.68 tuber per degree C temperature increase. As a base value is defined 12 tubers, at 14 degrees C average daily temperature. Using the regional set of meteorological and soil data POTATO-ZONING calculates the expected number of tubers.

Figure 9 gives an example of such calculation for Indonesia. When combined with yield data of the same region a good approximation of tuber size distribution can be made assuming a bell shaped size distribution around the average tuber weight. It is obvious that often the areas within a region with the highest yields also show the highest dry matter concentrations and tuber size.

Heat tolerance explorations

All explorations shown earlier were based on an average day/night temperature window for potato growth between 5 and 21 °C. Below this temperature it is too cold and above it too warm for adequate potato growth. POTATO-ZONING, however allows the exploration of cultivars with higher or lower heat tolerance while maintaining its low temperature tolerance level at 5 °C. When cultivars with increased heat tolerance are grown, potato

Figure 9. Trends in number of tubers per plant in Indonesia.

cultivation can be expanded into hotter areas and into hotter periods during the year. This indicates that the length of the suitable growing season increases in an area with a too hot summer for potato production. So a proper exploration should properly take into account the increased length of a growing season associated with the increase in tolerance. Figure 10 shows such explorations with a regional data set in South Asia with the production characteristic: the attainable temperature sum of the growing season of the first potato crop. To grow at least one crop, at least 1250 K d has to be accumulated within the 5-21 °C window. If the length of the growing season exceeds 2200 K d a second season starts if at least another 1250 K dwill be accumulated. Figure 9 shows the calculated distribution of the areas with at least three growing seasons and the length in K d of the third season in South Asia. The upper temperature tolerance limit is assumed to be 21 °C (standard), 24 °C and 27 °C average daily temperature, respectively, and the lengths of the growing seasons to be 2000, 2200 (standard) and 2400 K d. The lower the high temperature tolerance level and the longer the growing season lasts, the fewer growing seasons in a year and the shorter the third season lasts. Fresh yields based on temperature-related dry matter concentration often are of greater interest than the tuber dry matter yields shown here. For the fresh market a low dry matter concentration may be advantageous (selling water) but to store potatoes and especially for processing a high dry matter concentration is needed. At lower altitudes

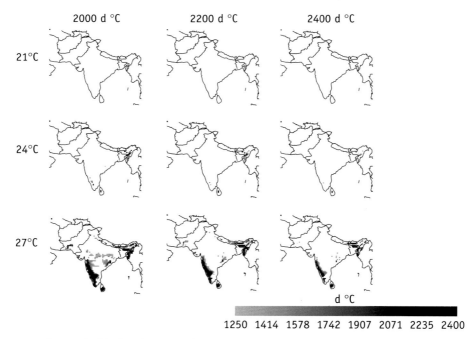

Figure 10. Area suitable for three growing seasons and length of the third growing season in K d at three levels of tolerance of maximum temperature and three imposed maximum lengths of a growing season.

in tropical regions, the dry matter concentration decreases with increasing temperatures until they become prohibitive. A spin-off of breeding efforts to increase dry matter concentrations would be an adaptation to lower lying production zones. Agro-ecological zoning is a tool to explore the impact of such efforts. Breeding for heat tolerance, increasing the temperature threshold, results in an increase in areas suitable for potato production. Increasing the temperature requirement results in a slight shift of the growing season; the crop needs more days to accumulate the required thermal time. The selection of the upper temperature threshold has a distinct effect on the spatial distribution of the growing seasons, and shows a strong increase of the number of growing seasons. It is concluded that breeding for heat-tolerant potato cultivars increases considerably the area suited to potato production.

Conclusions and perspectives for future development

Further explorations possible with POTATO-ZONING

Users of agro-ecological zoning studies are interested in several aspects of the potato production chain. The question 'Where to grow my potatoes?' can be answered by decision support zoning studies such as the present one. The studies reveal comparative advantages of one area over another. Do farmers need to use irrigation? How far is the production area from population centres? The studies will also show how long it takes for a crop to mature, and, depending on the length of the growing season and its average temperatures, how high the dry matter concentration of the tubers may be. The higher the concentration, the better the tubers are suited for storage and processing. Other feasible calculations, not shown in this chapter include the tuber size distribution. The tuber size distribution is closely related to the number of tubers per plant and the total fresh tuber matter per plant. Higher yields and lower numbers lead to increased grade sizes that can be derived from dry matter yield calculations (Figure 6), dry matter concentration calculations (Figure 8) and the tuber number per plant (Figure 9).
What is the impact of an irrigation scheme on yields of different crops? Calculation examples of the Indo Gangetic Plain in other studies by the authors (unpublished) have shown that potato is twice as efficient in producing tuber dry matter as wheat in producing grains. Wheat has a low harvest index because the crop has to invest in straw. Rice is one quarter as efficient as potato as it not only has a low harvest index but the paddy-grown crop also loses water through evaporation and then leaching is not even accounted for. Moreover, potato is grown in a cool season with a low evaporative demand, unlike rice, which is grown when temperatures are high. POTATO-ZONING can be used to calculate the water use efficiency in potato production by shifting the growing season to cooler periods of the year or by exploring the impact of shortening the season when it gets too hot late spring.
The zoning as shown in both the *global* and the *regional* studies can be used for risk analysis. When long-term weather data are available for many meteorological stations within an area it is possible to calculate where and how often a crop will be exposed

to hazards such as droughts, frost or excessive rains. It is possible also to explore the risks of changed cultural practices such as using later or earlier cultivars or earlier or later plantings.

With POTATO-ZONING the regional effects of climate change on potato production can be calculated. In such studies the presently prevailing temperatures can be increased by 1 or 2 degrees and the resulting shift in growing periods, hazards and yields can be calculated and compared with the present situation.

Not every area that is suitable for ware potato production can be used for the production of seed potatoes. Seed potatoes need a long frost free period but with low temperatures to reduce the development of bacteria and of aphids as vectors of viruses. Adequate maps will show where the thermal time increases relatively slowly, indicating a comparative advantage in such areas over others. The use of meteorological data will also indicate where certain pests are likely to become important such as Colorado beetle in hot and dry summers or certain diseases such as late blight in cool and rainy growing seasons

Further research and development needed

The methodology followed here is still amenable to improvement. The light use efficiency could be made dependent on the prevailing light intensity. Resource use efficiency decreases with the availability of the resource. *E.g.* ample water and nitrogen lead to low water use and nitrogen use efficiencies. Kooman & Haverkort (1995) found good relations between mean daily incident solar radiation and the light use efficiency. Incorporating daily minimum and maximum temperatures would increase the accuracy of the explorations compared to the average day and night temperatures. For real desert-like areas where field capacity is never reached, a threshold should be built in: *e.g.* no water-limited production possible when the value of (rainfall - evapotranspiration) is below a certain value.

In the LINTUL model used, it is assumed that a potato variety exists that is suitable for the growing season defined. The plants of that variety emerge, distribute their dry matter and die in a manner such that at the end of the growing season (because the maximum period is exceeded or because temperature become prohibitively high or low) the foliage is dead and 80 % of all the dry matter produced during the whole growing season ended up in the tubers. Such varieties may exist in real life but, even if they do, they may not necessarily meet the consumers' or processor's demands. Suitable varieties from a quality point of view may have lower yields but a higher economic return. Ideotyping studies are helpful to identify the suitability of a particular variety in a specific environment.

Improvement of the potato processing chain through quantitative approaches has much scope for further development. Together with partners in developing countries, the multinational food industry and funding agencies cases are being developed. From experience to date, several options for further research arise:

* Agro-ecological zoning for processing and seed potato in different regions, for instance Eastern Europe.

* Ecoregional study of potato in the Asian Potato-Cereal Belt (Pakistan, Nepal, India, Bangla Desh, China and Vietnam.

* Potato Processing in the Indo-Gangetic Plain in India can be explored. (There, near Calcutta, all potatoes are harvested in March and cause a glut on the market. It should be possible to use solar dehydration (it is hot and sunny in April and May) and to process potatoes for the local market.

POTATO-ZONING should be extended with relation between climatic data and more quality aspects than just dry matter concentration. Examples are susceptibility to black spot bruising and concentrations of reducing sugars.

Reduced grid size (higher resolution) is needed for most users, who are not interested in the global scope but are more interested in a particular country or region within a country. The availability of data is often a problem and it is also relatively costly to digitise maps and data them when they are not available in digital form.

References

FAO, 1990. FAOCLIM Global Weather Database. Agrometeorology group, Remote Sensing Centre, Research and Technology Division, FAO.

FAO, 1996. Digital soil map of the world and derived soil properties, version 3.5, November 1995, derived from the FAO/UNESCO soil map of the world, original scale 1:5 000 000. CDROM, FAO, Rome

Haverkort, A.J., 1990. Ecology of potato cropping systems in relation to latitude and altitude. Agricultural Systems 32: 251-272.

Haverkort, A.J. & P.M. Harris, 1987. A model for potato growth and yield under tropical highland conditions. Agricultural and Forest Meteorology 39: 271-282.

Hawkins, A., 1957. Highlights of a half-century in potato production. American Potato Journal, 343(1):25-29

Hijmans, R.J., 2001. Global distrbution of the potato crop. American Potato Journal 78: 403-412.

Keulen, H. van and W. Stol, 1995. Agroecological zonation for potato production. In: A.J. Haverkort & D.K.L. MacKerron (eds.), Potato ecology and modelling of crops under limiting growth. pp. 357-372.

Kooman, P.L. & A.J. Haverkort, 1995. Modelling development and growth of the potato crop as influenced by temperature and daylength: LINTUL-POTATO. In: A.J. Haverkort & D.K.L. MacKerron (Eds.) Potato ecology and modelling of crops under conditions limiting growth. Kluwer Academic Publishers Dordrecht, pp. 41-60.

Leemans, R. & W. Cramer, 1991. The IIASA database for mean monthly values of temperature, precipitation and cloudiness on a global terrestrial grid. Report RR-91-18. International Institute for Applied Systems Analysis (IIASA), Laxenburg, Austria.

Müller, M.J., 1987. Handbuch ausgewählter Klimastationen der Erde. 4. verbesserte Auflage. Forschungsstelle Bodenerosion Mertesdorf (Ruwertal), Universität Trier, Gerold Richter, Trier, 346 pp.

Müller, M.J., 1996. Handbuch ausgewählter Klimastationen der Erde. 5. ergäntzte und verbesserte Auflage. Forschungsstelle Bodenerosion Mertesdorf (Ruwertal), Universität Trier, Gerold Richter, Trier, 400 pp.

Rowe, R.C. (ed), 1993. Potato Health Management. APS Press. Minnesota, USA

Spitters, C.J.T., 1987. An anlysis of variation in yield among potato cultivars in terms of light absorption, light utilization and dry matter partitioning. Acta Horticulturae 214:71-84.

Stol, W., G.H.J. de Koning, P.L. Kooman, A.J. Haverkort, H. van Keulen & F.W.T. Penning de Vries, 1991. Agro-ecological characterization for potato production. A simulation study at the request of the International Potato Center (CIP), Lima, Peru. Report 155, CABO-DLO, Wageningen, 53 pp.

Verhagen, A., P.W.J. Uithol, C. Grashoff & A.J. Haverkort, 1998. Agro-ecological zoning of potato, worldwide potential and water-limited yields. AB note 151.

Name of the DSS

CropScan nitrogen decision support system

Who owns the DSS, Principal author's, address,...

Bert Meurs, Plant Research International, P.O. Box 16

6700 AA Wageningen, the Netherlands

bert.meurs@wur.nl

What is it for, what questions does it address?

How much nitrogen does a grower need to apply mid season to achieve the final tuber yield he or she is aiming for?

What is the suitable timeslot for this application?

How does the grower make sure that not unacceptable residual amounts of nitrogen stay in the soil after harvesting the crop.

What input is required?

Expected final tuber yield as it is closely quantitatively linked to the amount of nitrogen the standing biomass should contain well before harvest.

Crop reflection characteristics and their relationship with crop nitrogen content

Timeslot for application depending on variety lateness and expected time of harvest

Who is it for, who are the intended users?

Designed to be used by growers and advisors

What are the advantages over conventional methods (whatever those may be)

This method has two great advantages over the two other methods often used (soil sampling and crop sampling) to determine nitrate concentrations of the soil or of crop parts: this method is non destructive and gives the amount of nitrogen to be applied instantly.

How frequently should the DSS be used or its values be updated?

Once or twice per season is sufficient

What are the major limitations?

As for most supplemental nitrogen dressing decision support systems, the total amount needed to be applied depends on the final yield expected. hazards such as droughts, frost or epidemics cannot be predicted with accuracy so there is some uncertainty there.

What scientific / technical enhancements would be desirable?

The crop scan needs to be calibrated annually. A more robust apparatus is needed.

3. Crop-reflection-based DSS for supplemental nitrogen dressings in potato production

R. Booij[†] and D. Uenk
[†]Dr. Remmie Booij passed away on December 10, 2003.

To achieve a certain yield level a potato crop needs to amass sufficient nitrogen in its foliage well before crop maturity. This paper shows a decision support system for supplemental nitrogen dressing using the following quantitative relationships: the relationship between final tuber yield and minimum total foliage nitrogen content and the timeslot when this amount should be present. Crop reflection characteristics were related to this amount of nitrogen assuring that all elements of the decision support system are in place. A grower will need to know what yield he or she is aiming for and about half the required amount of nitrogen is then supplied before or at planting. Before mid-season the crop is assessed with crop reflection at two wavelengths to verify the amount of nitrogen in the standing biomass. If the difference between the amount present and required is applied.

Introduction

Nitrogen has been a hot issue in potato production for many years now. Firstly because the availability of nitrogen determines yield and quality and secondly, excess of nitrogen has a negative effect on the environment by pollution of the surface and ground water. To reduce the adverse effects of excessive nitrogen application, the nitrogen supply during crop growth should correspond with the crop nitrogen demand during the period of crop growth. Of course, an important prerequisite is the maintenance of yield (economic returns). This means that during crop growth the crop nitrogen status should be monitored and that supplementary nitrogen should be applied, based on relevant set points.

However, nitrogen availability in the soil is very dynamic, not only between years, but also within a growing season due to variability in mineralization, leaching, denitrification and crop nitrogen uptake (Blackmer, 1998). Therefore, care is needed in the timing and level of nitrogen application. Modelling studies by De Koeijer and Oomen. (1997) and by Booltink and Verhagen (1997) and experimental work of Stone *et al.* (1996) showed that significantly better results can be obtained if the nitrogen application rate is made time-dependent.

Monitoring crop nitrogen status is necessary if one is to obtain temporal precision in nitrogen application. The quality of light reflected from the canopy can tell us about the nitrogen status of the crop (Ma *et al.*, 1996; Stone *et al.*, 1996, Bausch and Duke,

1996; Heege and Reusch, 1997; Baret and Fourty, 1997; Vouillot *et al.*, 1998). The relation between crop nitrogen status and canopy light reflection in certain wave bands is based on the relations between nitrogen and leaf area development and between nitrogen concentration and chlorophyll concentration (Baret and Fourty, 1997). Crop light reflection, therefore, may be a tool to determine crop nitrogen status.

Plant Research International has developed a system, which is being applied currently in practice, in which the supplementary nitrogen application in potatoes is based on the measurement of crop light reflection.

The system

For the design of a system for supplementary nitrogen application for potatoes, the following aspects are essential:
- An easy method for determination of crop light reflection.
- A relation between crop light reflection characteristics and crop nitrogen status.
- Nitrogen uptake needed during crop growth to obtain a certain yield level (set points)

The equipment

A standard, handheld, multispectral radiometer (CropScan, Inc., Figure 1) was used to measure crop light reflection in 8 wavebands.

Light reflection and nitrogen status

The crop light reflection depends on the wavelength. Light reflection is particularly high in the near infrared region (Figure 2) and much lower in the visible region. Nitrogen

Figure 1. CropScan in use in a potato field.

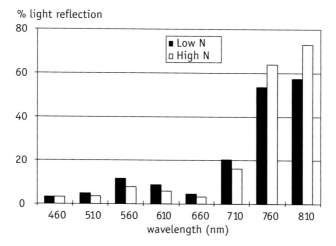

Figure 2. Effects of nitrogen availability on the percentage of light reflection at different wavelengths.

availability affects the crop light reflection pattern. At lower nitrogen availability a lower proportion of the light in the red and infrared region is reflected, and a higher proportion of that in the green and orange region (Figure 2). Based on these effects a reflection characteristic was developed, which related the reflection pattern to the nitrogen status of the crop. Figure 3 shows that the reflection characteristic of the potato crop is affected by the nitrogen application rate and crop development (or time). The reflection characteristic appears to be lower at lower nitrogen availabilities and decreases with crop development (Figure 3).

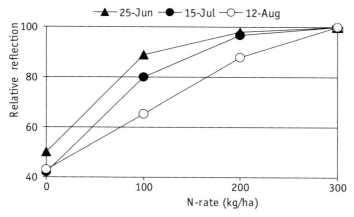

Figure 3. Time course of the relation between nitrogen availability and proportional light reflection characteristic (100% = value of characteristic at 300 kg N/ha).

Development of set points

Now we are able to determine the crop nitrogen status using crop light reflection, the next question is what should it be at crucial moments during crop growth.

Total nitrogen content (in the standing biomass) increases rapidly at the beginning of crop growth, reaches a maximum and remains constant or decreases slightly afterwards (Figure 4). The maximum nitrogen uptake in the Netherlands is reached at the end of June or the beginning of July. The amount of nitrogen in the crop at the time the nitrogen content has reached its maximum, roughly reflects the amount of nitrogen that is finally recovered in the tubers at harvest (Figure 4). Based on this information, set points for nitrogen content were defined.

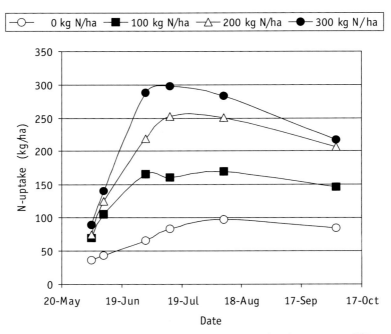

Figure 4. Pattern of nitrogen in a potato crop during crop development at different rates of nitrogen availability.

A test of an early prototype was carried out in a number of field experiments in which the amount of nitrogen at planting, the timing of determining requirements and application were varied. The results (Figure 5) showed that by using the system one could achieve a similar yield to the control, independent of the amount of nitrogen applied at planting and, further, that a time-window for application of approximately two weeks (difference between T1 and T2) around the beginning of July was found (Figure 5).

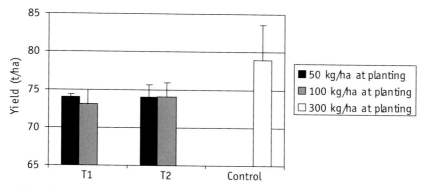

Figure 5. Performance of the system at different levels of nitrogen at planting (50 or 100 kg N/ha) and time of application (T1 or T2). Difference between T1 and T2 about two weeks around the beginning of July.

The basics of the system are now as follows:
- Apply a sub-optimal amount of nitrogen at planting (total mineral nitrogen availability at planting, including mineral soil nitrogen at planting should be around 150 kg N/ha).
- Measure the crop light reflection in the first week of July
- Determine crop nitrogen content and compare with the set point for the crop nitrogen content
- Apply the difference between actual content and the set point.

Validation in practice

The performance of the system, as described above, was tested in practice in 2000 on nine different potato fields for starch production. These fields were split into two parts. One part received the normally recommended rate of nitrogen, to be decided by the farmer, before planting. The other part received the equivalent of only 100 kg N /ha before planting. All the nitrogen that was applied before planting originated from organic manures. On the part of the field that had received only 100 kg N/ha at planting, crop light reflection was measured between July 1 and July 10 and supplementary nitrogen was applied, based on the relationships developed earlier.

The system for supplementary N application was compared with the standard system on four aspects, namely final yield, nitrogen input, nitrogen off-take and residual mineral soil nitrogen. The two approaches resulted in the same yields, nitrogen off-takes and residual mineral nitrogen (Figure 6). However, the total nitrogen input was about 20% less (Figure 6). So a similar yield was obtained at a lower nitrogen input. However, this difference was not reflected in the residual mineral soil nitrogen (Figure 6). As the nitrogen off-take was also the same (Figure 6), it means that the amount of nitrogen in the senesced foliage was less.

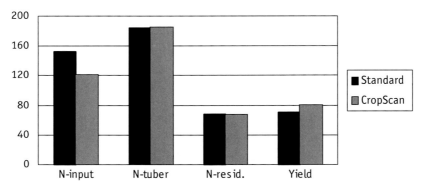

Figure 6. Comparison of the system for supplementary nitrogen application based on crop light reflection (CropScan) and the standard application (Standard) for effective nitrogen input, nitrogen offtake, residual nitrogen in the soil (0-60cm) and yield after harvest. Data are the means of nine fields.

Conclusion

Although, in the system being tested, the nitrogen applied at planting is sub-optimal for crop growth over the whole growing season, it should be sufficient to support growth until the supplementary nitrogen is applied. By splitting the nitrogen rate, in which the second application depends on crop nitrogen status, better use can be made of nitrogen from organic sources (manure, soil organic matter, crop residues, etc.) The set points for initial nitrogen rate and the supplementary nitrogen rate might depend on the soil type. However, to study this aspect is only worthwhile if the standard recommended rates depend on soil type.

For farmers it is difficult to accept that they have to "wait" until the beginning of July, before a recommendation can be given. From other systems (e.g. petiole sap), that are more or less diagnostic in nature, they are used to receive recommendations at any time during crop growth. However, it is questionable whether we need such a diagnostic tool during early potato growth. If sufficient nitrogen is applied at planting for the leaf canopy to reach 100% soil cover, the risk that nitrogen shortage would develop during that period is very low, because losses through leaching are unlikely to be high. The CropScan system for supplementary nitrogen application, combining a sub-optimal rate at planting and measuring as late as possible, has shown that it is possible to "wait".

As for all systems for supplementary N application the problem can be that the readings might be affected by other environmental factors (e.g. diseases, drought, etc.). This can also be the case with the CropScan readings, To overcome this problem a new approach is being developed, in which an internal field reference is used.

An important advantage of the CropScan system for supplementary nitrogen application compared with other systems, is the speed of the system. Instantaneously the nitrogen status can be measured and the supplementary nitrogen can be calculated, so that growers don't have to wait on the results from a laboratory.

References

Baret, F., & Th. Fourty, 1997. Radiometric estimates of nitrogen status of leaves and canopies. In: Diagnosis of the nitrogen status in crops, G Lemaire (Ed.), Berlin: Springer Verlag, pp 210-227.

Bausch, W.C. & H.R. Duke, 1996. Remote sensing of plant nitrogen status in corn. Transactions of the ASAE 39 1869-1875.

Blackmer, A.M., 1998. Using precision farming technologies to improve management of soil and fertilizer nitrogen. Australian Journal of Agricultural Science 49 555-564.

Booltink, H.W.G. & J. Verhagen, 1997. Using decision support systems to optimize barley management on spatial variable soil. In: Application of systems approaches at the field level, .M.J. Kropff *et al.* (Eds.), Dordrecht: Kluwer Academic Publishers, pp 219-233

De Koeijer, T.J. & G.J.M. Oomen, 1997. Environmental and economic effects of site specific and weather adapted nitrogen fertilization for a Dutch field crop rotation. In: Precision agriculture '97. Spatial variability in soil and crop, J.V. Stafford, (Edt.) Oxford: Bios, pp 379-386.

Heege, H.J. & S. Reusch, 1997. Zur teilflächenspezifischen Stickstoff-Kopfdüngung Landtechnik. 52 126-127.

Ma, B.L., M.J. Morrison & L.M. Dwyer, 1996. Canopy light reflectance and field greenness to assess nitrogen fertilization and yield of maize. Agronomy Journal 88 915-920.

Robert, P.C., R.H. Rust & W.E. Larsen, 1999. Proceedings of the fourth international conference on precision agriculture. American Society of Agronomy.

Vouillot, M.O., P. Huet & P. Boissard, 1998. Early detection of N deficiency in a wheat crop using physiological and radiometric methods. Agronomie 18 117-130.

Name of the DSS

AZOBIL for Potatoes

Who owns the DSS, principal author's, address,...

C. Chambenoit ALTERNATECH Agro-Transfert Domainde de Brunehaut, F-80200 Estrées Mons, France

F. Laurent ARVALIS-Institut du végétal, Station Expérimentale, F-91720 Boigneville, France

J.M. Machet INRA Unité d'agronomie Laon-Reims-Mons, F-02007 Laon cedex, France

What is it for, what questions does it address?

What is the optimal Nitrogen fertilizer rate for my potato crop?

What is the effect on Nitrogen uptake of a planting delay ? haulm killing delay?

What is the effect of soil types on mineralization from humus?

What input is required?

Potato crop planting and haulm killing date, soil information (soil C content and other chemical and physical soil characteristics), previous crop, fate of crop residues and frequency of organic manure application within rotation, previous cover crop, type and amount of recent N fertilization (organic manure), readily available soil mineral nitrogen at plantation, date of measurement

Who is it for, who are the intended users?

This decision support is intended for :

• potato growers to better manage fertilizer N supply

• potato processing industry concerned by tuber quality

• consultancy services that want to improve their recommendations

What are the advantages over conventional methods (whatever those may be)

The conventional method for N fertilizer management is generally to apply large amount of fertilizer once at plantation. Using AZOBIL for potatoes enable us to take in account soil mineral N supply and crop N requirement during the growth cycle duration. It has as advantages, at field specific scale, to better match N supply and crop N requirement, leading more closely to expected crop maturity and to reduce non desirable N losses to the environment.

How frequently should the DSS be used or its values be updated?

AZOBIL for potatoes is to be used annually for field specific assessment at plantation.

What are the major limitations?

AZOBIL for potatoes enables producers to manage tuber yield *and indirectly crop maturity and thus tuber quality*. But today questions of farmers concern directly the implementation of practices able to meet various market's quality requirements

What scientific / technical enhancements would be desirable?

AZOBIL for potatoes has to be adapted to take in account various market's quality requirements

Which nitrogen nutrition level enables potato to meet quality requirements ?

Which other factors governing physiological process of quality elaboration of potato ?

4. Development of a decision support system for nitrogen management on potatoes

C. Chambenoit, F. Laurent, J.M. Machet and H. Boizard

In this chapter, we review the state of the art in terms of new guidelines developed for the potato crop, allowing estimation of the optimal dose of nitrogen at the start of the season. The guidelines principally concern the evaluation of nitrogen requirements and soil supplies in accordance with production objectives and the crop cycle duration.

From 2001 on, we started to promote these new guidelines to French farmers and their advisers in the form of potato-specific tools and decision support systems:
- a guide to nitrogen fertilization of the potato
- technical notes and software for calculating doses

The entire set of established benchmarks will be integrated into a new version of Azobil currently under development.

Introduction

Nitrogen is a key factor in potato crop production: good nitrogen management simultaneously influences the crop's tuber quality, productivity and environmental impact. Faced with end-user demands, nitrogen management techniques will have to become more and more precise.

For the producer, managing the nitrogen fertilization of a potato crop means:
- giving the crop enough nitrogen to allow it to attain the best yield and the best tuber quality, depending on the plot's pedoclimatic conditions, the variety's potential and production constraints (work organization, specifications, etc.).
- reducing the post-harvesting nitrogen residuals in the soil and loss of nitrogen into the environment (notably via nitrate wash-out and emission of greenhouse gases), in order to meet demands for environmental protection.

Achieving this dual objective necessitates precise adjustment of nitrogen supplies to fit the plant's needs. In France, calculation of the total dose of nitrogen is based on the "forecast balance" method for mineral nitrogen in the soil:

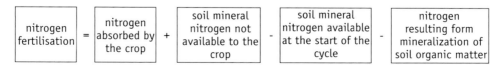

This method has been developed by the INRA for all the common annual crops, with the help of a decision support system called "Azobil" (Machet *et al.*, 1990).

Thanks to a study programme on the agrophysiology of the potato carried out between 1992 and 2001 and led by ALTERNATECH's AGRO-TRANSFERT section, the balance method has been improved by adapting it more specifically to potato crop production.

Evaluation of a potato crop's nitrogen requirement

Up until now, evaluation of a potato crop's nitrogen requirements was based on fixed-volume values. The objective of the work presented here was to enable production objectives to be taken into account, together with the duration of the potato crop's vegetative cycle.

Crops and nitrogen nutrition

Under satisfactory nitrogen nutrition conditions, a crop's accumulation of nitrogen is strongly linked to the development of its biomass. Examination of the relationship between this biomass and the nitrogen content enabled us to establish a diagnosis of the extent to which a stand's requirements are met (Lemaire *et al.*, 1989).

The work carried out by ALTERNATECH's AGRO-TRANSFERT section on the characterization of nitrogen requirements has led to definition of the potato crop's critical curve for nitrogen (Duchenne *et al.*, 1997) . This describes the greatest possible production of total dry matter (haulms + tubers) with the lowest possible nitrogen content (Figure 1).

Nitrogen Nutrition Index, NNI = %N measured / %Nct

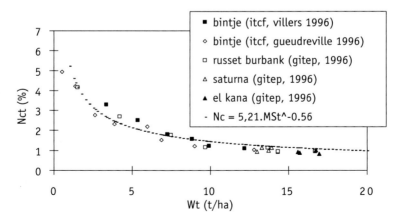

Figure 1. The potato crop's critical curve.

For each biomass level (Wt), we can now compare the measured nitrogen content with critical nitrogen content. We define the Nitrogen Nutrition Index (NNI) as the ratio of the measured nitrogen concentration to the critical nitrogen concentration. This NNI characterizes the crop's nitrogen nutritional status.

On the critical curve, the value of the NNI is 1. The crop has reached the optimum state of nitrogen nutrition for biomass production.

Below the curve, NNI is less than 1, and so N limits growth

Above the curve, the value of NNI is greater than 1, which is what we call "luxury consumption".

For the potato crop, the critical curve and the nitrogen nutrition index constitute a means of parameterizing the nitrogen requirement in terms of the production objective. The next question is how to determine which levels of NNI are optimal for production criteria other than total biomass - for example parameters related to the development of yield or tuber quality.

Optimal nitrogen nutrition levels

Initial work on characterising optimal nitrogen nutrition levels has dealt with the total yield for varieties destined for processing.

Figure 2 shows the relationship between the nitrogen nutrition index (NNI) measured just before haulm destruction and the yield measured at harvest. Figure 2. shows results of 24 trials carried out between 1991 and 2000. Cultivars: Bintje, Kaptah Vandel, Monalisa, Nicola, Russet Burbank. Soils studied: silt, sandy silt, silty clay and chalk.

In order to allow simultaneous analysis of results from a whole set of plots and years, the yield observed for a given dose of nitrogen has been expressed with respect to the potential yield in the absence of limiting nitrogen. The total yield index (YI) is equal to the ratio between the tuber yield and the maximal yield observed with a non-limiting dose (the plateau of the nitrogen response curve).

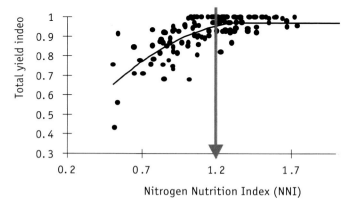

Figure 2. The nitrogen nutritional level needed for maximizing tuber yield.

The correlation between the pre-haulm-destruction NNI and the YI is significant. This confirms the links between nitrogen absorption and the development of yield, and thus the pertinence of the NNI for forecasting the achievable value yield with respect to maximum potential yield. The optimum level of NNI for maximising the total yield is always greater than 1 and about 1.2.
The critical curve and the NNI enable us to calculate crop N uptake

N uptake = %N . Wt
Critical curve: %N = NNI . %Nct

N uptake = NNI . %Nct . Wt
The NNI characterises the nitrogen nutritional level
%Nct characterises the critical nitrogen content
Wt, the dry matter weight of the plant as a whole, characterises the growth level

We can thus adapt the evaluation of crop N requirement to:
• production objectives, by using the appropriate crop N status (NNI) and
• the production duration cycle, by estimating dry matter production (Wt)
The dry matter of the plant as a whole can be simulated by a growth model.

Optimal crop N requirement

For each production area, the nitrogen requirement was evaluated with the help of a crop model using:
• Monteith's concepts (1972 ,1977) for simulating the growth in biomass
• the critical curve (Duchenne *et al.*, 1997) for evaluating the nitrogen requirement corresponding to each biomass level (the CRITIC model by T. Duchenne)
The CRITIC model (Figure 3) enables simulation of the crop's nitrogen requirement as a function of the crop cycle duration, the production zone's climate and the level of nitrogen nutrition decided by the user, according to the production objective.

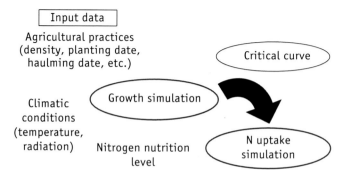

Figure 3. General organization of the CRITIC model.

The model functions with a one-day time unit and simulates the potential total production of dry matter (i.e. under non-limiting conditions) between the date of emergence and that of haulm-destruction. The model requires few input data, most of which are readily available on commercial farms: climatic data (radiance, temperature) and crop characteristics (distance between rows, planting density).

To allow practical use of this model, it was necessary to evaluate the accuracy of our simulations (Figure 4).

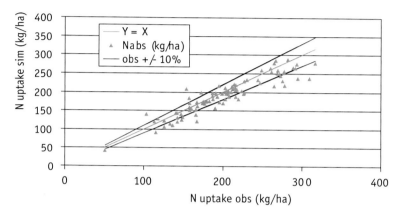

Figure 4. Evaluating the simulation of N uptake 14 experiments, with a wide range of soil nitrogen availabilities, cultivars (10), years (5) and N fertilizer rates (from 0 kg/ha to 300 kg/ha).

The figure shows simulations of all the situations studied. There is very good agreement between the observed and predicted nitrogen uptake values, even in situations where a crop nitrogen deficiency occurred.

These evaluation results encouraged us to propose this model for estimating crop nitrogen requirements.

In each production zone, the frequency-domain use of the model over 30 climate years with an NNI level of 1.2 allowed us to determine potential production and the corresponding nitrogen requirement. The production level selected for generating recommendations was the maximum value observed at least 8 years out of 10, meaning that requirements were satisfied in 4 cases out of 5.

The new nitrogen requirement guidelines were publicized in the form of a guideline table (Table 1). The nitrogen requirement thus determined allows farmers to achieve the maximum production potential for a given pedoclimatic context and vegetative cycle: it replaces the "fixed-volume requirement" used up until now.

Taking haulm destruction dates into account in the evaluation of nitrogen requirements allows us to integrate the earliness of a given variety.

For example, the requirement of an early potato planted on the 5th of April with haulms destroyed on the 31st of July is estimated at 190 kg/ha. For the same plantation date,

Table 1. Nitrogen absorbed by the crop (kgN/ha) - Production zone 1.

Planting date	Haulm destruction date						
	21/07 - 31/07	01/08 - 10/08	11/08 - 20/08	21/08 - 31/08	01/09 - 10/09	11/09 - 20/09	21/09 - 30/09
21/03 - 31/03	190	205	215	225	225	230	230
01/04 - 10/04	190	200	210	215	220	225	230
11/04 - 20/04	190	195	210	215	220	225	230
21/04 - 30/04	185	195	205	210	220	225	230
01/05 - 10/05	175	190	200	210	215	220	225
11/05 - 20/05	165	180	195	205	210	215	220
21/05 - 31/05	155	165	180	195	200	205	210
01/06 - 10/06	130	155	170	185	190	200	205

the requirement of a late potato with haulms destroyed on the 10th of September is estimated at 220 kg/ha.

These new nitrogen requirement guidelines also allow a farmer to adapt to a given year's conditions by taking the planting date into account.

For example, in the case of a firm-fleshed variety with haulms destroyed on the 31st of July, late planting on the 11th of May gives a nitrogen requirement value of 165 kg/ha versus 190 kg/ha for earlier planting (5th of April).

During wet springs (with a vegetative cycle shortened by very late planting), the use of these new guidelines can thus enable more precise adjustment of nitrogen supplies to the cycle length and thus an improvement in the crop's maturity at the time of haulm destruction, the tuber quality and the preservation characteristics during storage, whilst avoided an excess of nitrogen.

These nitrogen requirement guidelines are available to producers throughout France - temperature and radiance conditions are defined for each production zone.

Evaluation of soil nitrogen supply for potato crops

The goal of this work was to better take account of the potato crop's particularities (rooting behaviour, cycle length etc.) in the evaluation of soil nitrogen supply.

Soil mineral nitrogen available at the start of the cycle

The crop's use of soil mineral nitrogen depends on root depth but also the abundance and the spatial distribution of the various soil horizons. Knowledge of rooting behaviour now enables us to open up pathways for improving the evaluation of available nitrogen in each horizon. Soil type, structural state, irrigation and even the variety thus appear to be pertinent criteria for determining the quantity of soil mineral nitrogen available to the crop at the start of the cycle.

We do not yet possess experimental results which validate the utility of taking these factors into account when estimating soil nitrogen supplies. Nevertheless, surveys carried out in the Picardy region of France have shown that the exploitation of nitrogen by roots in the 30-60 cm horizon is always at least 50%. Farmers are thus encouraged to measure the soil nitrogen reservoir at the beginning of the cycle in the first 60 cm (horizon 1 = 0-30 cm and horizon 2 = 30-60 cm) and to only count half of the nitrogen content in the second horizon.

The date of measurement of the N reservoir at the start of the cycle corresponds to the opening of the nitrogen balance. This date must be as close as possible to the planting date. If this measurement is carried out well before planting, the nitrogen "stock" evaluated could be very different to that available at the start of the growth phase (i.e. soon after planting). Early performance of the soil nitrogen measurement leads to poor estimations of soil nitrogen reserves. If heavy rainfall occurs between the reservoir measurement and the provision of nitrogen, the decrease in available mineral nitrogen must be re-evaluated in order to take wash-out into account.

Nitrogen generated by the mineralization of soil humus

Evaluation of soil humus mineralization now takes into account the date of measurement of the mineral nitrogen reservoir prior to plantation and the date of haulm destruction. This mineralization is evaluated on the basis of the quantity of potentially mineralizable nitrogen over a year, which in turn depends on the type of soil. (Table 2).

The quantity of mineralizable nitrogen is a potential and annual value that needs to be adjusted to fit each situation. An initial adjustment takes into account any policy for returning organic matter to the soil (fate of harvest residues, frequency of organic fertilization) by applying a weighting coefficient (Table 3). This coefficient serves to

Table 2. Quantity of potentially mineralizable nitrogen (kgN/ha).

OM	CaCO3 <100			100 < CaCO3 < 400			Ca CO3 >400
	Clay <120	Clay 120-230	Clay 230-450	Clay <120	Clay 120-230	Clay 230-450	Clay <120
<10	35	20	15	30	20	10	15
10-15	85	55	35	70	45	30	45
15-20	120	80	50	100	65	40	60
20-25	155	100	65	125	80	50	80
25-30	190	120	75	155	100	60	95

Table 3. Coefficient linked to policy for returning organic matter to the soil.

Harvest residues	Frequency of organic fertilization			
	Nothing	5-10 years	3-5 years	< 3 years
Removed or burned	0.8	0.9	1.0	1.1
Ploughed in every second time	0.9	1.0	1.1	1.2
Always ploughed in	1.0	1.0	1.2	1.3

take into account the existence of an "active" fraction of organic matter, fed by significant additions of fresh organic matter.

In addition, the plant will more or less use all the total mineralizable nitrogen, depending on the duration of vegetative growth. The quantity of nitrogen effectively useable by the plant is that mineralized between the date of reservoir measurement (near to the time of planting) and the date on which the haulms are destroyed. A second adjustment takes into account the real duration of the production cycle via a "duration of mineralization factor" (Table 4).

Soil mineral nitrogen not available to the crop

Not all of the soil mineral nitrogen is not available to the potato crop: the mineral nitrogen residual mineral nitrogen measured at the end of the crop cycle corresponds to the quantity of soil nitrogen that is not useable.

Table 4. The duration of mineralization factor.

Date of haulm destruction	Date of residual measurement						
	01/03 - 10/03	11/03 - 20/03	21/03 - 31/03	01/04 - 10/04	11/04 - 20/04	21/04 - 31/04	01/05 - 10/05
01/07 - 10/07	0.45	0.40	0.35	0.35	0.30	0.25	0.20
11/07 - 20/07	0.50	0.45	0.40	0.35	0.35	0.30	0.25
21/07 - 31/07	0.50	0.50	0.45	0.40	0.35	0.35	0.30
01/08 - 10/08	0.55	0.50	0.50	0.45	0.40	0.35	0.35
11/08 - 20/08	0.60	0.55	0.50	0.50	0.45	0.40	0.35
21/08 - 31/08	0.65	0.60	0.55	0.50	0.50	0.45	0.40
01/09 - 10/09	0.65	0.65	0.60	0.55	0.50	0.50	0.45
11/09 - 20/09	0.70	0.65	0.65	0.60	0.55	0.50	0.50

Up until now, this residual N was estimated at harvest according to the type of soil and by considering that all crops had the same rooting efficiency. Current knowledge calls into question the harvest-time evaluation of this residual N - in terms of the significance of the level found in the soil and the date on which the measurement is performed.

In the case of the potato crop, the period during which nitrogen is absorbed by the plant ceases shortly after the haulms are destroyed, i.e. when all the vegetative parts are dead. Closing the nitrogen balance for potatoes must therefore be performed at the time of complete senescence of the stand (or at the time of haulm destruction if this occurs first) and not on harvest. Values for soil mineral nitrogen not available to the crop are as follows (Table 5):

Table 5. soil mineral nitrogen not available to the crop (kgN/ha).

Soil type	
Sandy and silt soil	20
Clay and chalk soil	40

Harvest-time measurement of the residual N must nevertheless be performed. In fact, mineralization of organic nitrogen continues between haulm destruction and harvest. In the case of the potato, the harvest-time residual resembles an environmental marker. Indeed, a high harvest-time residual constitutes a threat to the environment, as the nitrogen is likely to be washed out into the soil profile during the winter. When a high harvest residual is either suspected (crop management error, poor yields etc.) or measured, farmers are advised to plant an intermediate "nitrate trap" crop which will limit winter wash-out. Mineralization of the residue from the intermediate crop should be taken into account when calculating the total dose of nitrogen for the subsequent potato crop.

With respect to the mineralization of other sources of organic matter (crop residues, organic amendments etc.), please refer to the article by Goffart which appears elsewhere in this book.

Validation of the results

Our database included 124 nitrogen experiments. The main factors of variation were the cultivar (12), the year (10) and the N fertilizer rate.

For each nitrogen data point, we assessed crop N uptake, soil inorganic N content (before planting and after harvesting) and soil N mineralization. The nitrogen concentration in the plant as a whole was measured prior to senescence.

For each experiment, we determined the optimal level of N fertilization for yield maximization.

Our improvements were evaluated by comparing the rates calculated using the novel and the traditional methods respectively, with the optimal N rates for each experiment (Figure 5).

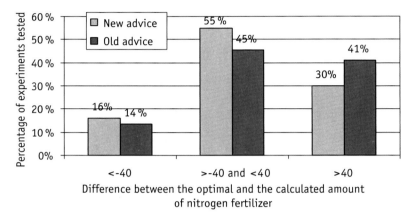

Figure 5. Evaluation of advice on 44 experiments where the optimal amount of nitrogen fertilizer was reliably determined.

Our novel, calculated N rate enables evaluation of optimal levels of nitrogen fertilization with greater accuracy than the traditional approach (55% vs. 45 %). The NNI is thus a very efficient way of determining the nitrogen nutrition level for maximal tuber yield. The novel, calculated N rate also reduces cases of over-fertilization (30% vs. 41%) but not cases of under-fertilization (16% vs. 14%). In our database, under-fertilization cases appear when there is a high risk of wash-out (sandy soil, high levels of mineral N before planting, etc.). In these cases, division of the total amount of nitrogen into several amendments is advised.

Decision support systems developed in France for potato growers

Transfer of the above results and guidelines to potato producers and agricultural advisers was performed by promoting methods and decision support systems for calculating the total dose of nitrogen.

Methods for evaluating the total dose of nitrogen were notably spread via technical information days and test platforms. These events are systematically accompanied by the publication of articles in the farming press. A special issue of the monthly publication "Perspectives agricoles" was published in March 2003, with five articles focusing on the nitrogen fertilization of potatoes. In order to ensure the transfer of recommendations, handy tools for calculating the total dose of nitrogen for potatoes have been developed. A computer program for calculating the total dose of nitrogen has been made available to producers and advisers The various nitrogen balance items are entered via a system of tabbed menus (Figure 6).

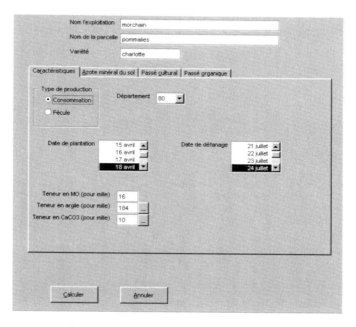

Figure 6. Computer screen of the DSS' tabbed menus.

An A4-format output sheet sums up all the entry data and displays the values calculated for the various balance items, as well as the calculated, total dose of nitrogen. Furthermore, in each region, technical information sheets for calculating the total dose of nitrogen are made available to producers and advisers (Figure 7.)

Our results as a whole have been made available as part of a practical guide (Chambenoit *et al.*, 2002) to the nitrogen fertilization of potatoes published jointly by Agro Transfert, ITCF and INRA in January 2002. This manual shows how scientific knowledge works in practice and is designed to assist producers. All guidelines are explained in a simple and easily understandable way. The guide is intended for agricultural consultants and farmers but is also recommended for students and teachers.

Figure 7. *Technical information sheets for calculating the total dose of nitrogen are made available to producers and advisers.*

Conclusions

In terms of estimating nitrogen requirements, use of the CRITIC model demonstrates that coupling the critical curve with a crop growth model enables us to determine at any given moment the quantity of nitrogen necessary and sufficient for achieving the growth potential.

To extend the model still further, more work is needed to incorporate not just yield considerations but also quality objectives into the crop models. This new way of calculating nitrogen fertilization has been promoted to potato producers in France, and enables them to match the total dose of nitrogen to the actual length of the crop cycle. Estimation of requirements between planting and haulm destruction now takes into account the earliness of the variety. From now on, evaluation of mineralization will take into account the date when the soil N reservoir is measured and that of haulm destruction. The total dose of nitrogen calculated therefore limits cases of over-fertilization (and thus maturity problems at the time of haulm destruction) without endangering yields. The set of results acquired for potatoes and presented here will be integrated into the new version of Azobil, which is due to be released in 2004 under the name "Azofert".

Acknowledgements

The authors thank Mr T. Duchenne for his valuable work contributions.

References

Chambenoit, C., F. Laurent, J.M. Machet & O. Scheurer, 2002. Fertilisation azotée de la pomme de terre - guide pratique. Published by Agro-Transfert/ITCF-ITPT/INRA. 130 pages

Duchenne, T., J.M. Machet & M. Martin, 1997. Diagnosis of a potato nitrogen status. In : G. Lemaire (Eds.) Diagnosis of the itrogen status in crops. Springer, Berlin, 119-130.

Lemaire, G, F. Gastal & J. Salette, 1989. Analysis of the effect of N nutrition on dry matter yield and optimum N content. XVI International Grassland Congress, Nice, France, 179-180.

Machet, J.M., P. Dubrulle & P. Louis, 1990. AZOBIL : a computer program for fertilizer N recommendation based on a predictive balance sheet method. Proceedings of first congress of the European society of agronomy

Montheith, J.L., 1972. Solar radiation and productivity in tropical ecosystems. Journal of Applied Ecology, 9 : 747-766.

Monteith, J.L., 1977. Climate and efficience of crop production in Britain. Phil. Trans. Soc., London, 281 : 277-294.

Name of the DSS

Potato N fertilization strategy

Who owns the DSS, Principal author's, address,...

GOFFART Jean-Pierre, OLIVIER Marguerite

Crop Production Department

Agricultural Research Center of Gembloux

4 rue du Bordia, 5030 Gembloux, Belgium

goffart@cra.wallonie.be; olivier@cra.wallonie.be

What questions does it address?

- what will be the field-specific N-advice for my potato crop at planting time?
- how can I manage the splitting of this advised level of N?
- how can I detect the crop's need for supplemental-N during the growing season?
- if required, when do I have to supplement fertilizer-N during the growing season?

What input is needed?

- Part A of the strategy (field specific N-advice assessment): potato crop type and variety, soil information (soil C content and other chemical and physical soil characteristics), previous crop, fate of crop residues and frequency of application of organic manure within rotation, previous cover crop, type and amount of recent N-fertilization (organic manures as well as mineral supply), readily available soil mineral nitrogen at planting, mean annual temperature.
- Part B of the strategy (assessment of crop N status): field specific N-advice value, date of crop full emergence, in-season chlorophyll-meter measurements, variety coefficient for determination of chlorophyll-meter threshold value.

Who is it for, who are the intended users?

This decision support system is intended for:

- potato growers better to manage fertilizer-N supply during the growing season
- potato processing industry concerned about the expected tuber quality
- consultancy services that want to improve their recommendations for potato growers

What are the advantages over conventional methods (whatever those may be)

The conventional method for N fertilizer management is generally to apply a large amount of fertilizer once at planting. This does not take into account real annual variability in supply of soil mineral N and in the crop's requirement for N. Using the POTATO N FERTILIZATION STRATEGY has advantages, at field specific scale, better to match N-supply and crop N-requirement, leading more closely to expected tuber quality and to reduce undesirable losses of N to the environment.

How frequently should the DSS be used or its values be updated?

POTATO N FERTILIZATION STRATEGY is to be used annually for field specific assessment from planting to up to 55 days after crop emergence.

What are the major limitations?

The DSS needs a variety-specific coefficient for use with the chlorophyll-meter.

What scientific / technical enhancements would be desirable?

High efficiency of supplemental N-fertilizer should be guaranteed

5. Management of N-fertilization of the potato crop using total N-advice software and in-season chlorophyll-meter measurements

J.P. Goffart and M. Olivier

With the aim of matching at field scale the potato crop's N requirements with mineral nitrogen supply from soil and fertilizers, the splitting of total N fertilizer application has been combined with in-season assessments of crop N-requirements and indirect soil mineral nitrogen supply. After assessment of a total N-advice based on the predictive balance-sheet method (software Azobil, INRA, Laon, France) and before planting, 70% of the advised amount is applied to the crop at planting. Subsequently, between 25 and 55 days after crop emergence, the need for supplemental N is assessed through non-invasive and quick, in-season measurements of leaf chlorophyll-concentration directly within the field. A simple conditional relation has been established to support a potato grower's decision on application of the remaining 30% nitrogen. Its application requires chlorophyll-meter measurements (with the Hydro N-tester model) within the fertilized field and within a small 'window' without any applied N-fertilizer (zero-window). The strategy developed is economically feasible and easy to operate. It also gives the grower the possibility of saving on N-fertilizer and of reducing the risk of excessive N fertilization and its negative impact on tuber quality and environment

Introduction

The objective of the work was to develop a complete strategy to guide the producer in the management of the nitrogen fertilization of the potato crop at a field scale and based on split N-applications. The basic idea is that establishment of N-advice for the crop at planting can never be accurate as it is not possible to predict with precision total crop N-requirements and the soil mineral N supply that will occur during the season. Splitting the N fertilization combined with the follow up of estimating the crop N-requirement during the growing season and, indirectly, the soil mineral N supply can help to solve this problem (Vos & MacKerron, 2000). The strategy developed includes five steps as shown in Figure 1. Step 1 aims to determine at field level a total N-advice for the crop before planting and based on the predictive balance-sheet method as recently summarised by Vos & MacKerron (2000). The software used for this purpose is "Azobil" developed at the Institut National de la Recherche Agronomique (INRA, Laon, France) by Machet *et al.* (1990). In step 2, N is applied at planting at 70% of the N-advice rate. In fields with a low level of soil mineral N supply especially, field trials have indicated that it is not advisable to reduce the N applied at planting by more than 30% (Goffart *et al.*, 2002). Steps 1 and 2 (part A of the strategy) are implemented before the crop is planted. The three following steps (part B of the strategy) occur during the growing season. In step 3, the crop is monitored to determine its nitrogen status, that is, whether it has sufficient nitrogen or is deficient in it, according to the establishment of a threshold value for the measured parameter. Due to its very close relation with the leaf N concentration (Vos & Bom, 1993), the leaf chlorophyll concentration has been chosen for monitoring. A hand-held chlorophyll-meter apparatus enables quick non-invasive measurements on individual leaves in the field. In our case, we have developed this step of the strategy with the Hydro N-tester model of chlorophyll-meter (Hydro-Agri Europe, Brussels, Belgium), an improved version of the previous SPAD - 502 model (Minolta, Osaka, Japan). In step 4, the result of the measurements leads to the decision whether or not to apply the remaining 30% of the advised N according to the estimated need for it in the crop. Finally, in step 5, when supplemental nitrogen is required, the producer has to decide on the type and mode of nitrogen fertilizer to use to ensure the best efficiency of the supplement to the crop.

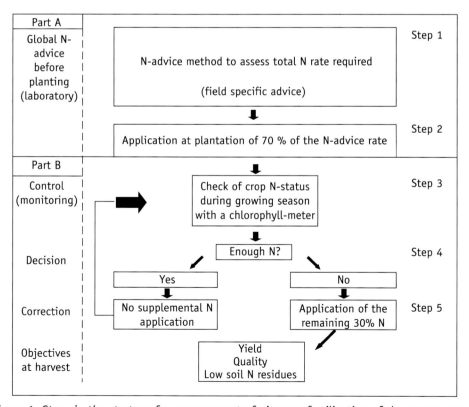

Figure 1. Steps in the strategy for management of nitrogen fertilization of the potato crop.

Quantitative basis of the balance-sheet method as developed in the Azobil software

The basic theoretical relation of the balance-sheet method at field scale as developed in Azobil is illustrated in Figure 2, and can also be written as (all the terms of this relation are expressed in kg N ha^{-1})

$$B + Rf = (Ri-L) + Mn + X$$

Where,

B is the crop's N-requirement, a fixed value according to the potato type (values used in Belgium are: 250 kg N ha^{-1} for ware (consumption) potatoes harvested at maturity, 200 kg N ha^{-1} for ware potatoes harvested before maturity, 160 kg N ha^{-1} for seed potatoes, 270 kg N ha^{-1} for starch potatoes although this type is hardly represented in Belgium). Such values assume insurance in nitrogen availability to reach maximal tuber yield for most of the potato cultivars (Chambenoit *et al.*, 2002).

Rf is the residual soil mineral nitrogen content at harvest, an arbitrarily fixed value according to the rooting depth (30 kg N ha^{-1} for a maximal potential rooting depth up to 90 cm as generally recognised for the potato crop in deep soil profiles).

Ri is the readily available soil mineral nitrogen in a determined depth of soil at the end of winter period (before planting). This value is assessed through soil sampling and analysis.

L is the soil mineral nitrogen potential losses that should occur due to nitrate leaching during the period from analysis of soil N to N-fertilizer application. It is an assessed value based on the Burns leaching model (Burns, 1975) which considers soil and sub-soil types and depth of appearance of the latest. It leads to a specific table of potential N losses for individual situations according to the number of rainy days and intensity of rainfall.

Mn is the net supply of soil mineral nitrogen during the growing season. "Mn" results from the sum of four individual components :

$$MN = Mh + Mr + Ma + Ap$$

Where
Mh is the net mineralization from soil organic matter
Mr is the mineral N supply from previous crop residues
Ma is the mineral N supply from organic manures (farmyard, slurry,) and catch crops
Ap is the mineral N supply from previous grass crop

All these four components are assessed from parameter values (for Mh, Mr, Ma and Ap) or from an algorithmic relation for Mh. Parameters are accessible to an expert software user (scientist or qualified technician) in order to adapt it to local conditions when data are available. Several parameters developed in the North of France have been adapted

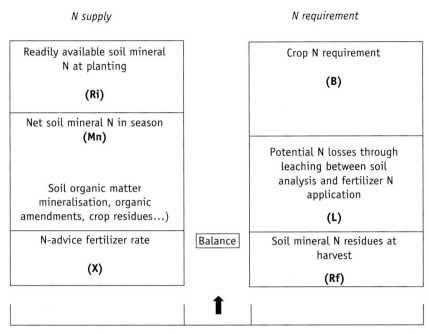

Figure 2. Scheme of the basic theoretical relation of the balance-sheet method at field scale as developed in Azobil software (version 1.2; Machet et al., 1995).

to Belgian conditions. As an example, details for Mh assessment in Azobil are described in the next section and other detailed parameters are described in Machet *et al.* (1995). **X** is the N-advice, corresponding to the total mineral N-fertilizer rate to be applied to the crop.

The current version of the model includes some important assumptions:
- different N sources are supposed to have the same recovery coefficient (100%)
- the model is additive, which means that no interactions are considered between the different terms of the balance-sheet
- gaseous N losses (volatilization, denitrification) are counterbalanced through N supply from the air and non-symbiotic fixation
- no leaching occurs under cropped soil
- main soil profile explored by the roots of the crop is fixed at 60 cm for the potato crop
- the start and end for the balance-sheet are at times characterized by relative stability of mineral N in soil - start at the end of the winter period (1st of March) and end at harvest time.

The model takes account of soil mineral nitrogen (nitrate and ammonium ions) for N supply.

Detailed descriptions

Here follow the detailed descriptions of the relations and parameters for assessment of the term "Mh" in the predictive balance-sheet method "Azobil"

Mh = Qn x CRF x COS

Where
QN is the amount of soil organic N mineralized
CRF is a coefficient related to the fate of crop residues in the rotation and to the frequency of application of organic manures within the field
COS is a coefficient for cropping period during the mineralization period which is fixed at a maximum value of 270 days corresponding to COS = 1 (from 1st of March to 1st of November). For instance, for a ware potato mid-late cultivar COS = 0.7 (= 190 days), with lower and upper limits for Mh fixed at 50 and 100 kg N ha^{-1}, respectively.

Details on assessment of **QN**

QN = MASSOL x 0.5 (MO/10) x {(1200 FC) / ((A + 200) (0.3 C +200))}

Where
MASSOL is the soil mass (tonnes/ha)
MO is the soil organic matter concentration (‰)
C is the CaCO3 soil concentration (‰)
A is the clay content (‰)
FC is a climatic factor corresponding to (0.2 T - 1) with T as local mean annual temperature (Celsius degrees)

MASSOL = 100 x PL x DA x 1.3 (1-CC)

Where
PL is the ploughing depth (cm) with a default value of 27 cm
DA is the soil apparent density, with a default value of 1.5 g/cm^3
CC is the concentration of stones in the soil (%), with a default value of 0 %
1.3 is a factor to allow for the fact that mineralization occurs within the ploughing depth and also in the neighbouring sub-soil (one third of the ploughing depth)

Details of CRF values are shown in Table 1

Quantitative basis for the assessment of potato crop nitrogen status through chlorophyll-meter measurements

The plant nitrogen concentration in the potato crop is closely related to the chlorophyll concentration in its leaves (Vos & Bom, 1993). The assessment of potato leaf

Table 1. CRF, the coefficient related to the fate of crop residues in the rotation and to the frequency of organic manures application within the field.

CRF values	Frequency of organic manures application			
Fate of crop residues	none	5 to 10 years	3 to 5 years	less than 3 years
Exported or burned	0.8	0.9	1.0	1.1
Ploughed every two years	0.9	1.0	1.1	1.2
Ploughed	1.0	1.1	1.2	1.3

chlorophyll concentration through quick chlorophyll-meter (HNT) measurements has been developed in field trials conducted on Belgian loam soil from 1997 to 2001 (Goffart et al., 2002).

Optimal conditions for use of HNT to assess crop nitrogen status have been determined (see following section). Within a crop early in the season, absolute HNT values are only discriminative when comparing fertilized and non-fertilized plots (Goffart et al., 2002; Denuit et al., 2002). Cultivars, locations and soil water availability are the other most important factors influencing absolute chlorophyll-meter values. It is recommended that no further measurements are made after supplemental N fertilizer has been applied.

Relative threshold values for HNT measurements are then required to assess the crop nitrogen status of the crop, and finally to support the grower's decision to add supplemental N to the crop. Relative HNT threshold values, specific to potato cultivars and suppressing the location effect that is present with absolute HNT values, were determined according to the following relation (Olivier & Goffart, 2002; Goffart et al., 2002):

$$(HNT_{field} - HNT_o) > N_{field} \times C$$

Where
HNT$_{field}$ is the HNT values measured within the field fertilized with 70% of the advice
HNT$_0$ is the HNT values measured within a small plot located inside the field and in which no N was applied ("zero N - window" as reference plot)
N$_{field}$ is the N rate (70% N-advice) in kg N ha^{-1} applied to the field
C is a cultivar specific coefficient (for instance for cv. Bintje, C = 0.5. C values for other cultivars are still being established).

This relation sets the condition for the decision to apply supplemental N (the remaining 30% N). In this relation, HNT$_0$ integrates information on soil mineral nitrogen supply, while HNT$_{field}$ tells of the state of development of the fertilized crop and its potential during the growing season according to the nitrogen readily taken up by the crop. Neither piece of information is accurately predictable at planting time but it is essential that each should be assessed more precisely so as to improve the total rate of

N-application according to the crop's real requirement for N. The relation was developed in unirrigated conditions, but is however only usable when water-supply does not limiting crop development and growth at any time during the growing season. It also requires the setting up of a small plot of about 200 square metres in which no N is applied (zero N-window) within a homogeneous part of the field. The 'window' is easily produced by switching off the fertilizer sprayer over a determined distance while applying nitrogen.

The zero-window has to be localised in a homogeneous and representative part of the field. If the field includes heterogeneous areas, each one should be considered individually for HNT measurements and each should include a specific zero-window.

Description of the decision support system in practice

To run the decision support system at field level, the required equipment is the software "AZOBIL" and an HNT chlorophyll-meter.

Part A of the strategy

The necessary data for AZOBIL are listed as input data in Table 2. All these data are concerned with field history and practices which should normally be available from the potato grower.

When visiting a grower who is asking for N-advice, an intermediary belonging to an advisory laboratory collects the field data on a standard form, and samples the soil in the field in question, from 0 to 30 cm and from 30 to 60 cm depths for analysis of soil mineral nitrogen and also for organic matter in the 0 to 30 cm soil layer.

The information and the results of soil analysis are computerized with the software through an easy-to-fill window screen. Then the N-advice is immediately assessed and the outputs are given either on the computer screen or printed on paper.

The Azobil outputs comprise three forms:

Form 1 shows the grower's address and the location of the field, a graphic representation of the soil mineral nitrogen profile and the total fertilizer N-advice with eventual comments from the advisory laboratory (Figure 3).

Form 2 summarises the field soil characteristics, field history, previous crop characteristics, and recent nitrogen applications (organic as well as mineral)

Form 3 gives the mean annual temperature, N balance-sheet with detailed information (Ri, Mh, Mr, Ma, Ap, Ma), fertilizer N-advice again and a table of potential N-losses through leaching before application of N-fertilizer. This table enables the grower to increase the N-application above that advised to match any N estimated to be lost following recorded rainfall during the period.

These three sheets are sent to the grower for information and advice before the crop is planted.

Table 2. *Detailed field information required as input data for the establishment of the N-advice through the predictive balance-sheet method, Azobil.*

Nature of the field information	Detailed information	Balance-sheet elements affected
• Grower information	Address, location of field, field name	--
• Potato crop type	Consumption, seed or starch potato + cultivar	B
• Soil information	– Clay, sand and CaCO$_3$ concentration in the ploughed layer	Mh
	– Organic matter concentration in the ploughed layer	Mh
	– Potential rooting depth of the crop	L, Ri, Rf
	– Sub-soil type	L
	– Depth of sub-soil appearance	L
	– Stones concentration	L, Mh
	– Ploughing depth	Mh
• Field history	– Age of previous grass crop	Ap
	– Date of ploughing of previous grass crop	Ap
	– Fate of crop residues	Mh
	– Frequency of organic manure application	Mh
• Previous crop	*(to identify the effect of previous crop residues)*	Mr
• Recent nitrogen fertilization	– Mineral nitrogen fertilizer recently applied (type and amount) aiming to deduce it from initial X value	X value
	– Last organic manure application (type, amount and date)	Ma
	– Previous cover cropping (species, biomass production and ploughing date)	Ma
• Available soil Nmin at planting	*(measurement made as close as possible to but before the date of planting)*	Ri
• Mean annual temperature	*(local mean annual temperature available from official services)*	Mh

(1) see correspondence with Figure 2 and related text

The cost of the software AZOBIL is about € 1250 Euros and the software is licensed and commercialised by INRA (Laon, France). The use of the software is generally assumed by an advisory laboratory. The cost of the soil analyses for mineral nitrogen and organic matter content is about € 25 per field, which can be met by the grower.

Part B of the strategy

To be able to assess the crop nitrogen status during the growing season, chlorophyll-meter measurements have to be done in both field and zero N-window areas, starting 25 days after crop emergence. The crop emergence must then be followed and complete

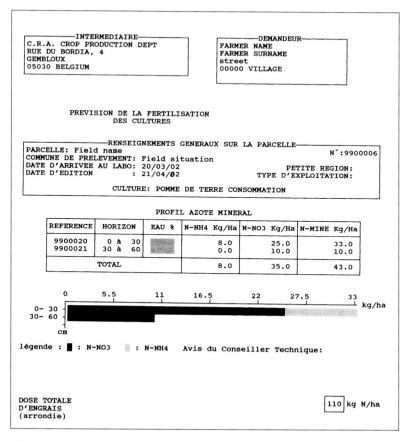

Figure 3. Illustration of screen shots for Azobil outputs - Form 1 of 3.

emergence is taken to be when about 75% of the plants have emerged. Measurements have to be done at 5- to 7-day intervals between 25 and 55 days after crop emergence, on the distal leaflet of the 4th or 5th leaf of a main stem, counting from the top of a plant (Figure 4).

Using the HNT chlorophyll-meter (Figure 5), 30 individual measurements are required to get a mean representative value for the field being analysed.

A "W" pathway pattern is recommended across a homogeneous area of the field being monitored. The area should not exceed 3 hectares. Along this pathway, at least 2 series of measurements (2 times 30 individual measurements) have to be done in the fertilized field. One series is enough in the zero-window. The detailed procedure is given by Olivier *et al.* (2001) and Goffart *et al.* (2002). The task is easy with the HNT so that the job can be done either by personnel of advisory laboratory or by the grower himself. A field form is supplied so that the user can collect the necessary information.

Figure 4. Representations of the latest completely developed leaf (picture on the left) and the leaflet of this leaf (picture on the right) used for a single measurement with the HNT chlorophyll-meter.

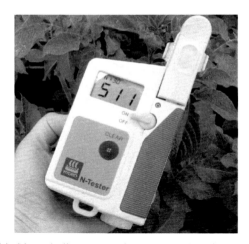

Figure 5. The hand-held chlorophyll-meter Hydro N-tester (HNT).

Figure 6 gives a representation of this form with experimental values as examples, and determination of the date when the need for the supplemental nitrogen fertilizer is detected.

The HNT chlorophyll-meter is commercially available and costs about 1500 euros (exclusive of VAT). The cost could limit its purchase by individual growers and so hamper the widespread adoption of the method. However, it can also be bought by a small syndicate of growers, and also it can be used on other crops such as winter wheat (Denuit *et al.*, 2002). The measurements can also be made by service laboratories.

Field name:............................ **Potato cultivar:**........................ **Year:**..............

Total N-advice (kg Nha^{-1}) (1)	140
N field (70% of (1)) (N rate applied at planting)	98
HNT threshold value (= N field x C*)	49 (for cv. Bintje)
Date of 75% crop emergence	**May 20th**

D.A.E.	Calendar date	HNT$_{field}$ indice**	HNT$_0$ indice***	HNT$_{field}$ - HNT$_0$	Decision****
25	June 14	622	594	28	No suppl. N
30	June 19	601	565	36	No suppl. N
35	June 24	580	535	45	No suppl. N
40	July 4	546	482	64	30% Supplemental N applied
45	July 9	-	-	-	-
50	June 14	-	-	-	-

*   C = Cultivar Coefficient for HNT threshold value assessment
    «C» values proposed for different potato cultivars:
    Bintje: 0,5 – Agria: 0,5 –Astérix: 0,7 – Nicola: 0,6 – Charlotte: 0,5 – Franceline: 0,6
**   HNT$_{field}$ indice is the mean value of at least 2 series of 30 individual measurements
***   HNT$_0$ indice is the mean value of 1 series of 30 individual measurements
**** Condition for supplemental N application: (HNT$_{field}$ - HNT$_0$) > (HNT threshold value)
    D.A.E..= Days After Emergence - Recommended dates for measurement at 5 to 7 days interval

Figure 6. Standard form for the follow up of the potato crop N status through measurements with the HNT chlorophyll-meter and for a decision on the need for supplemental N, with practical values as example for cv. Bintje.

Moreover, the cropped potato field area required to compensate for the cost of the method is not so great. *In the case where supplemental N is not required and not applied* following HNT measurements, it has been demonstrated on the basis of the mean cost of N-fertilizers (Ammonium Nitrate or Urea) and the mean price for tubers that 9 to 18 hectares of potato crop are sufficient to compensate for the cost of the measurement, through savings on N-fertilizer for an Azobil N-advice ranging from 200 and 100 kg N ha^{-1}, respectively (Goffart *et al.*, 2002).

Where supplemental N is required, a small increase in yield has generally been observed due to good timing of the supplementary application of N. Then, the potato crop area required to compensate for the cost of measurement compared to the cost of the whole amount of N being applied all-at-once at planting ranges between 1.3 and 7.2 ha (Goffart *et al.*, 2002)

Experiences with the strategy developed

The strategy has been studied in field trials from 1997 to 2000, and was at the development and validation stages on a commercial field scale in 2001 and 2002. Table 3 summarises the validation of the strategy for cultivars Bintje and Saturna in 2001.

Out of 9 locations on Belgian loam soils, supplemental N was required in 5 (nos. 2, 3, 4, 8 and 9). The decision to apply the supplement or not was right for 7 of the 9 fields. The decision was validated taking into account effects on yield, tuber size, tuber dry matter, tuber nitrate content and level of soil mineral nitrogen residues at harvest. In the two remaining locations (nos. 6 and 7) for which the decision was wrong, either the crop was irrigated (no. 6) or was planted to another potato cultivar (no. 7), suggesting

Table 3. Validation of the strategy for management of the N fertilization of a potato crop including the decision on the need for supplemental-N based on in-season HNT chlorophyll-meter measurements. Demonstration trials in 9 commercial fields on Belgian loam soil in 2001.

Location		Assessment of the need for N supplement	Decision on the need for supplemental N with the HNT test	Correctness of the decision	Date of the decision when supplemental N required (DAE*)
Number	Potato cultivar				
1	Bintje	no	No suppl. N	OK	-
2	Bintje	yes	Suppl. N	OK	38
3	Bintje	yes	Suppl. N	OK	37
4	Bintje	yes	Suppl. N	OK	32
5	Bintje	no	No suppl. N	OK	-
6	*Bintje + irrigation*	no	Suppl. N	Not correct	-
7	*Saturna*	no	Suppl. N	Not correct	-
8	Bintje	yes	Suppl. N	OK	27
9	Bintje	yes	Suppl. N	OK	38

Unirrigated crops except in location no. 6.
*DAE: Days After Emergence

the need for specific relative HNT values for named cultivars and for irrigated crop conditions. The need for supplemental N was detected early, between 27 and 38 days after emergence.

The decision support system was adopted successfully by 10 and 20 growers in a demonstration project in 2001 and 2002, respectively. The project was jointly developed in the Walloon Region with the potato association "The Walloon Potato Chain" (FIWAP, Gembloux, Belgium). Technicians of the association carried out the HNT measurements and training sessions were organized for technicians and growers. A descriptive folder of the DSS was distributed to the Belgian potato chain partners. Up to now, the Azobil software together with the local parameters adapted to Belgian conditions, is used exclusively at the Département Production Végétale, Gembloux which acts as reference advisory laboratory for total N-advice. Further development is planned for the use of the Azobil software in sub-regional laboratories in Belgium.

Growers find the strategy interesting because of the possibility of reducing the rates of N-fertilizer. Other listed advantages are: - better management of tuber characteristics such as size and dry matter concentration; - better quantitative assessment of the crop's N-requirement; - reduction in potato canopy development and consequently in risks of foliar diseases such as late blight in wet-seasons; - reduction in the risk of delayed crop maturity which can also affect the tuber quality and haulm destruction.

The next steps in the development of the strategy will be the training either of technicians from existing service and advisory laboratories or of growers in Belgium. More practical information on types and use of N-fertilizer for supplemental N application is also requested.

Prospects

Through a better assessment of the in-season N-requirements of the crop and of the soil mineral N supply, the strategy developed clearly gives the grower the opportunity to save N-fertilizer and to avoid the negative impact of excessive N on tuber quality and on the environment.

However, a first improvement of the method should be to improve further the accuracy of the Azobil N-advice. In potato trials conducted during recent years in Belgium and France with increasing rates of N, the variation of the advice around the calculated optimal rate is about 25 to 30 % (Goffart *et al.*, 2002). Current research work in France is now aimed at reducing this through a better assessment of total requirements for N and of in-season supply of soil mineral N from organic matter (Chambenoit, 2002). A new version of AZOBIL is under construction, taking into account N-losses through denitrification and volatilization (Machet *et al.*, 2001).

Another improvement of the strategy should be to modulate the present reduction of 30% of the N-advice according to the readily available soil mineral nitrogen at planting. However this must be done carefully in order to avoid limited efficiency of the supplemental-N for a crop that has suffered from insufficient N early in the growing season.

Instead of an arbitrary and single application of the remaining 30% N as supplemental nitrogen, this rate should also be modulated following the possibility of assessing more accurately the required supplemental nitrogen using the chlorophyll-meter measurements.

At present the relative HNT threshold values are determined for the cultivar Bintje in unirrigated conditions as this is the most frequent situation encountered in the Belgian potato region. Equivalent values still have to be established for other important cultivars and for irrigated conditions.

The application protocol for the supplemental N-fertilizer should also be subjected to more thorough investigation. Presently in Belgium, as irrigation only represents 4 to 5% of the potato cropping area, it is suggested that foliar urea should be applied at the maximum rate of 15 kg N ha^{-1} per application (Goffart *et al.*, 2002). Successive applications made at intervals of at least 5 to 7 days are recommended until the total N requirement is met, and they can take place simultaneously with the application of fungicide to control late blight.

Finally, this strategy for N-management should become well integrated in a more global DSS developed for the potato crop. A prerequisite for its use in other regions or countries should be the validation or adaptation of the parameters involved in the Azobil software to local climatic and soil conditions, and also the validation or adaptation of HNT relative thresholds values to locally-grown potato cultivars.

References

Burns, I.G., 1975. An equation to predict the leaching of surface applied nitrate. Journal of Agricultural Science, 85, 443-454.

Chambenoit, C., F. Laurent, J.M. Machet & H. Boizard, 2002. New evaluation of fertilizer N rates applied to potato crop. EAPR Abstracts of Conference Papers and Posters. 15th Triennial Conference of the European Association for Potato Research, Hamburg, Germany, 43.

Chambenoit, C., F. Laurent, J.M. Machet & O. Scheurer, 2002. Fertilisation azotée de la pomme de terre - Guide Pratique. Eds : ITCF/ITPT - INRA - AGRO-TRANSFERT, France, pp 128

Denuit, J.P., M. Olivier, M.J. Goffaux, J.L. Herman, J.P. Goffart, J.P. Destain & M. Frankinet, 2002. Management of nitrogen fertilization of winter wheat and potato crops using the chlorophyll meter for crop nitrogen status assessment. Agronomie, 22, 847-853.

Goffart, J.P., M. Olivier, J.P. Destain & M. Frankinet, 2002. Stratégie de gestion de la fertilisation azotée de la pomme de terre de consommation. Eds " Centre de Recherches Agronomiques de Gembloux", Gembloux, Belgium, pp 118. (web site: www.cra.wallonie.be)

Machet, J.M., P. Dubrulle & P. Louis, 1990. AZOBIL: a computer program for fertilizer N recommendations based on a predictive balance sheet method. Proceedings of 1st Congress of the European Society of Agronomy.

Machet, J.M., P. Dubrulle & P. Louis, 1995. AZOBIL: Manuel d'utilisation agronomique du logiciel. Eds: Station d'analyses agricoles de Laon et Institut National de la Recherche Agronomique (INRA, Laon, France). Financement Chambres d'Agriculture du Nord, du Pas-de-Calais et de la Somme; Agence de l'eau Artois Picardie, pp 33.

Machet, J.M., P. Dubrulle, N. Damay & S. Recous, 2001. Azofert: a decision support tool for fertilizer recommendations based on a new version of the predictive balance-sheet method. In "Book of abstracts - 11th Nitrogen Workshop", 9-12 September 2001, Reims, France, Books of abstracts, Eds INRA, 487-488.

Olivier, M., J.P. Goffart & M. Frankinet, 2001. Two methods of quick assessment of potato crop nitrogen status: working frame and threshold values leading to N-fertilizer supplement during growing period - 11th Nitrogen Workshop", 9-12 September 2001, Reims, France, Books of abstracts, Eds INRA, 503-504.

Olivier, M. & J.P. Goffart, 2002. Chlorophyll meter used as decision tool to manage nitrogen fertilization in potato crop. EAPR Abstracts of Conference Papers and Posters. 15th Triennial Conference of the European Association for Potato Research, Hamburg, Germany, 68.

Vos, J. & M. Bom, 1993. Hand-held chlorophyll meter: a promising tool to assess the nitrogen status of potato foliage. Potato Research, 36, 301-308.

Vos, J. & D.K.L. MacKerron, 2000. Basic concepts of the management of supply of nitrogen and water in potato production. In: Management of nitrogen and water in potato production, A.J. Haverkort & D.K.L. MacKerron (Eds.), Wageningen Pers, Wageningen, the Netherlands, 2000, pp 15-33.

Name of the DSS

The Potato Calculator

Who owns the DSS, Principal author's, address,...

Peter Jamieson, NZ Institute for Crop & Food Research Ltd, PB 4704, Christchurch, New Zealand

What is it for, what questions does it address?

- The potential yield at a site for a cultivar
- Water and N applications and timing to achieve the potential
- Effects of changing timing and amounts of water and N
- Amount of water drained below the profile, amount of N leached

What input is required?

- Soil description at least to root depth
- Weather data to date (radiation, temperature, precipitation) and scenarios of future weather
- Cultivar information
- Mineral N profile at or close to planting

Who is it for, who are the intended users?

Designed to be used by growers and advisors

What are the advantages over conventional methods (whatever those may be)

The system is predictive rather than reactive, and can respond to changes in recent weather and anticipated weather.

How frequently should the DSS be used or its values be updated?

Can be updated daily, but weekly or monthly should be sufficient

What are the major limitations?

We have limited cultivar information. Soil details should be obtained locally

What scientific / technical enhancements would be desirable?

Predictions of size distributions, effects of nutrients other than N, prediction of disease impacts...

6. Implementation and testing of the Potato Calculator, a decision support system for nitrogen and irrigation management

P.D. Jamieson, P.J. Stone, R.F. Zyskowski, S. Sinton and R.J. Martin

The Potato Calculator is a water and nitrogen management tool based on a mechanistic simulation model of potato growth. It uses the simulation model, records of recent weather and scenarios of future weather to estimate when water and N will become limiting. On the basis of this information, it creates a management schedule for irrigation and N-fertilizer applications. The schedule may be altered by the user at any time during the growth of the crop. Although the model runs simulations on default cultivar settings, the user may modify inputs in accordance with observed crop performance, by supplying emergence date, date of full canopy and average tuber number per plant. The Calculator is designed to replace crop analysis and monitoring systems for N management, so no other crop monitoring is necessary to operate the system. The model grows a canopy in response to temperature, with limitations associated with water stress and N supply. Light interception by the canopy drives biomass accumulation, and a simple biomass-partitioning rule initially directs all new biomass to the canopy, and eventually all to tubers. Nitrogen economy is based on assumptions about the N content of various organs. A major assumption is that tubers have a minimum requirement for N that is satisfied by taking N from stem storage; when stem storage is exhausted and no soil N is available, leaves die prematurely to supply N to growing tubers. Canopy size and biomass accumulation are modified by an index of water stress, i.e., the ratio of maximum water supply rate in the root zone to a demand rate based on potential evaporation and canopy size. Tests of the model and its implementation on farms are discussed.

Introduction

At 1.5% N of dry biomass, a potato crop of 90 t ha^{-1} at 23% dry matter content will contain 310 kg N ha^{-1}. Because most soils are unable to supply this amount of soil N during the life of a potato crop, fertilizer N must be added to make up the shortfall. How to calculate the shortfall and answer the question "when should N be supplied and how much?" is the aim of fertilizer decision support systems (DSS), be they paper, monitoring or computer based. They seek to supply N fertilizer in sufficient amounts and sufficiently early to avoid damaging deficits, but not in such large amounts to create a significant risk that the N will be lost to leaching before it is all taken up. A variety of methods have been used, many of them based on monitoring the N status of the crop (Gardner and Jones, 1975; Kleinkopf *et al.*, 1984, Williams and Maier, 1990), some relationship between aspects of biomass accumulation, N uptake and yield (Allison *et al.* 1999; MacKerron *et al.* 1993). Some of these methods have been shown to have doubtful validity (MacKerron *et al.*, 1995). Methods based purely on sap or tissue analysis may provide information on the optimum timing of N fertilizer application but, by themselves, give no indication of the optimum rate. However, the major problem with any monitoring system is that it is reactive; it cannot anticipate. We believe that the rules governing potato growth and development are now well enough known, that simulation models, including those describing soil processes, are mature enough, and the network of automatic weather stations is sufficiently well developed that the combination may be used to anticipate shortages of N, and management may be adjusted accordingly. Our confidence is reinforced by the success of a recently released wheat decision support system (DSS), the Sirius Wheat Calculator (Jamieson *et al.*, 2003), now in its second year of on-farm application.

Potato crops are also sensitive to water deficits (Jamieson, 1985a, Martin *et al.*, 1992) but can access water quite deep in the soil profile (Jamieson, 1985b). In a climate such as Canterbury, New Zealand, where summer evapotranspiration substantially exceeds precipitation, irrigation is essential for successful commercial potato production. Methods for budgeting water are well developed (Jamieson and Genet, 1985). The rules are simple and easily included in a computer program.

In this paper we describe a potato DSS, the Potato Calculator, and the potato simulation model at its core. The DSS is loosely based on and shares much code with the Sirius Wheat Calculator. The model simulates the growth and development of the crop, together with uptake and redistribution of N, and consequences for biomass accumulation and tuber growth. It uses the simulations, together with some user options for application amounts, to create a schedule of N and irrigation applications. The system is designed to run in real time, using constantly updated weather files and selected scenarios of future weather.

The model

The Potato Calculator simulation model produces a canopy (characterised by biomass and green leaf area index. (GAI) that intercepts light, and uses it to produce biomass at a light-use-efficiency (LUE) that is constant except under moderately severe water stress. The biomass is divided amongst leaves, stems and tubers according to simple partitioning rules. The potato model is based broadly on the Sirius wheat model (Jamieson *et al.*, 1998, Jamieson and Semenov, 2000), with which it shares the same soil description, percolation, leaching and evapotranspiration models. Obviously the models differ in detail because wheat grain grows only during canopy senescence, whereas potato tubers grow during canopy expansion as well as senescence. Otherwise, the potential growth model is, in principle, very similar to that of MacKerron and Waister (1985) except for the addition of the N economy of the plant. This uses the same approach as that of Jamieson and Semenov (2000), based on the role of N in the control of GAI (Grindley, 1997). The main simplifying assumptions are that specific leaf N is constant, that the stems contain a reservoir of labile N between upper and lower N concentrations, that structural N is a constant proportion of new biomass and is not retrievable, and that tubers have a minimum demand for N that they will take as they grow, even if this is at the expense of GAI expansion or the acceleration of senescence. The growth model can be written as:

$$Y = \int_{em}^{t} \xi A Q dt$$

where Y is the yield of tubers at time t, A is the LUE, Q is the daily light interception, ξ is a partitioning coefficient, and em is the value of t at emergence. Clearly this requires sub-models describing A, Q, ξ, and the time course of events. These are outlined below.

Canopy and phenological development

Phenological development in the model is both directly and indirectly associated with canopy development. The phenology model describes the lifetime progress in terms of thermal time between events such as planting, emergence, beginning of tuber growth, maximum GAI and completion of canopy senescence. The potential durations of these phases are assigned thermal durations from empirical determinations, and we expect these to vary for different cultivars. In the calculator, it is possible to modify the predictions by entering the observed dates of these events to keep the simulations close to reality.

Canopy expansion is calculated as a linear function of thermal time from emergence. The calculation is initially per plant, but the effect of population is brought in by having expansion cease when GAI (an area based determination) reaches a maximum assigned for a cultivar. For instance, the expansion rate for Russet Burbank is set at 12.5 cm^2 °C^{-1}day^{-1} per plant (Jamieson and Sinton, unpublished), giving a rate of increase of GAI of 0.005 °C^{-1}day^{-1} for a plant population of 44,444 ha^{-1}. Once the upper limit GAI value (6.0 for Russet Burbank) has been reached, GAI reduces as green area senesces at a set rate in thermal time (age related). The rate can be accelerated, either because of re-translocation of N to tubers, or because of water stress. During the expansion phase, the minimum requirements of N for structure and tuber growth must be met before the N for the GAI is met. Hence, if there is less N than this available, the canopy expansion is limited by the available N. Light interception is assumed to follow Beer's law with an extinction coefficient of 0.6.

The link between phenological and canopy development during stress is that the canopy ceases expansion at the time it would have reached its maximum value in the absence of stress; it will not "catch-up" later. There is also a link here between phenology and population because canopy expansion is calculated per plant, but the maximum is calculated per unit ground area. So reduced plant populations result in delays in reaching maximum GAI and increases in total duration. Hence, yield may not be much affected by variations in population.

Biomass partitioning

In the newly emerged crop, all new biomass is assigned to the canopy. The partitioning coefficient ξ initially has a value of zero. From the beginning of tuber growth, its value is set at 0.75. At the time a canopy growing without restriction would reach its maximum GAI, the value changes to 1. This means that during canopy expansion before tuber initiation all biomass is assigned to the canopy, and after maximum GAI all biomass is assigned to tubers. Canopy biomass is divided between leaf and stem by assuming a constant specific leaf weight of 35 g m^{-2} (Vos and van der Putten, 1998), with residual biomass assigned to stems.

N partitioning

N is partitioned according to a hierarchy of needs within the crop. Once tuber growth has begun, first priority for N goes to tubers to maintain their minimum N requirement of 0.8% (Martin *et al.*, 2001a) by mass. Hence, demand for N by tubers can cause premature senescence of the canopy by withdrawing N from green tissue. The next priority is for structural N in the canopy, assumed to be 0.5% by mass (Biemond and Vos, 1992). Following this comes the N required for leaf expansion at 2.0 g m^{-2} (Vos and van der Putten, 1998). Excess N goes into a labile pool that consists of nitrate stored in leaves (up to 0.4 g m^{-2}, Vos and van der Putten, 1998) and N compounds in stems (up to 4% by mass, Martin *et al.*, 2001a). Demands are filled by uptake of soil N if any is available, and then in turn from the lowest in the hierarchy to the step one below the current demander. Once maximum GAI is reached, the canopy becomes a net source of N, structural N is fixed, and canopy sources of N are run down as N is transferred to the tubers (Figure 1).

Fibrous root growth

The fibrous root system is described only by its depth in the profile. This is similar in principle to the wheat root model in Sirius. Roots extend downwards at a constant rate in thermal time to a lower limit. Although potatoes can extract water from at least a metre (Jamieson 1985b), their ability to extract resources from deep in the soil appears very restricted, leading to yield responses at small soil water deficits (Jamieson, 1985a), and varies substantially with cultivar. Accordingly we have set maximum rooting depth as a cultivar parameter, and restricted it to 0.7 m for Russet Burbank.

Evapotranspiration and water stress

Transpiration and soil evaporation are calculated separately as in Sirius. Daily transpiration demand is calculated from the modified Penman model of Ritchie (1972).

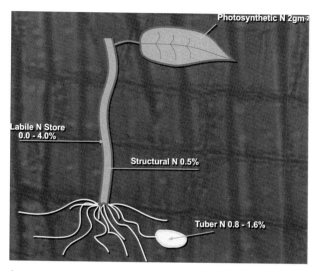

Figure 1. Assumptions on N content of various parts of the potato plant.

Daily supply is limited to 10% of the water held above the extractable lower limit (wilting point) in the root zone. A water stress index is calculated as the ratio of supply rate to demand rate, constrained to be unity or less. This is used to limit leaf expansion rates and accelerate canopy senescence.

Tuber moisture content

Tuber dry matter content is assumed to increase in thermal time according to a curvilinear model from 5% at tuber initiation to around 25% at 2000 °C days.

Experimental validation

Detailed validation of the model used data collected in an experiment at Lincoln, near Christchurch, New Zealand, performed in the summer of 1999/2000 (Martin *et al.* 2001a). Soil chemical determinations before planting showed there was 105 kg N ha^{-1} in the upper 0.5 m of soil. The crop was planted on 10 October 1999 (spring) with sufficient compound fertilizer such that nutrients other than N were not limiting. Among the treatments were four in which the timing and amount of N fertilizer was varied. The rates were 0, 150 and 300 kg N ha^{-1}, with the last one either applied all early (planting and ridging), or half applied early and the balance as a succession of 12.5 kg ha^{-1} applications at weekly intervals from 10 January. In that case only 275 kg N ha^{-1} was applied because the canopy had already died when the last two applications were due. The crop was irrigated seven times with applications totalling 260 mm. Rainfall during the season was 307 mm. Perversely, applications of 38 to 50 mm were often followed within 24 hours by significant rainfall events. On two occasions during December and early January this resulted in the crop receiving 75 - 100 mm of water in less than a week. Weather data were taken from a weather station approximately 2 km from the experiment. Measurements of GAI (Figure 2), biomass in tubers, leaves and stems (Figure 3), and their N content, were measured at approximately 14-day intervals throughout growth.

Less detailed data were available from experiments with winter-grown potatoes from Pukekohe, near Auckland, New Zealand (Martin *et al.*, 2001b). There data were confined to final yields. Again the major management variation was N supply, with applications of 0, 242, 350 and 472 kg N ha^{-1}. Weather data came from a weather station close to the site. Yield data were also available from experiments from Hastings, on the North

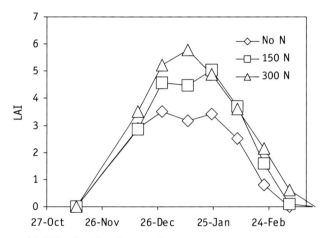

Figure 2. Observed LAI in the Lincoln experiment.

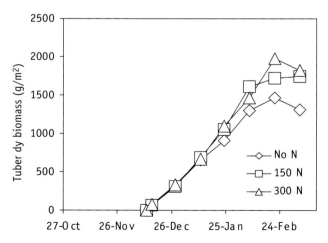

Figure 3. Observed tuber biomass accumulation in the Lincoln experiment.

Island East Coast (Stone *et al.*, 2003), where crops of Russet Burbank were grown over two years with adequate fertilizer and N supply (Figure 4). The main management variation was planting date, with plantings in November and December of 1998, and October and November of 1999.

Figure 4. Sites of validation experiments in Figure 6.

Results and discussion

Yields at Lincoln varied from 59.4 to 87.3 t ha^{-1} fresh weight. Simulated yields at the extremes were 57.3 and 85.6 t ha^{-1}. There was a close match between the observed and simulated time-courses of GAI, biomass and N accumulation in tissue for the highest and lowest N treatments at Lincoln (Figure 5).

In the three experiments reported, measured yield varied from less than 13 to nearly 100 t ha^{-1}. Generally, agreement between measured and simulated yield in response to variations in sowing date, season and N supply were good (Figure 6).

Much of the support from industry and Government for the Calculator project was given because these groups saw the calculator as a method of N-accounting by which they could establish that they were good stewards of the land, and not major contributors of nitrate pollution in groundwater. We used measurements of N leaching from Martin *et al.* (2001b) to test the accuracy of leaching calculations. The results were encouraging (Table 1), although leaching was underestimated in all but the highest N treatment. The discrepancies between the measurements and simulations suggest we have more work to do, either in improving our soil descriptions, or in parameterising the model of the processes.

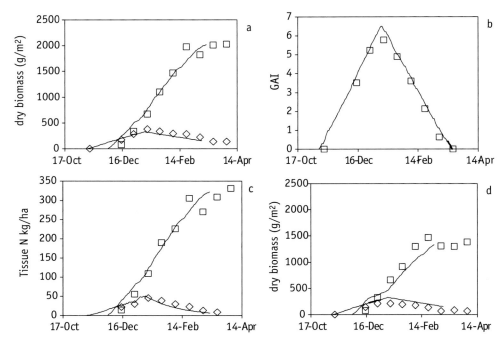

Figure 5. Time courses of biomass (a), GAI (b) and tissue N accumulation (c) for the highest fertilizer treatment and biomass (d) for the nil N (d) treatment at Lincoln. Points are observations, with squares in a, c and d referring to tubers and diamonds to tops. Lines are from the simulations.

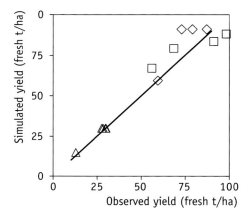

Figure 6. Comparison of observed yield with that predicted by the Potato Calculator simulation model. Data are from Lincoln (diamonds), Hawkes Bay (squares) and Pukekohe (triangles). The line is y = x.

Table 1. Comparison of measured and simulated leaching from the Pukekohe experiment. Amounts are in kg N ha^{-1}.

N Applied	Measured leaching	Simulated leaching
0	82	42
242	167	116
350	219	136
472	208	215

The calculator

It is clear from the foregoing that the simulation model at the heart of the calculator responds to management variations in much the same way that the crop does. It has the potential, using scenarios of future weather, to anticipate the effects of depletion of N in the whole system, and therefore to schedule the application of appropriate quantities of N-fertilizer before there is any plant response.

To do this, the calculator (Figure 7) first runs two simulations using the weather data available to it - real weather to the end of the current weather file, and a weather scenario from then on. One of the simulations is run assuming there is no limitation on N and water ("potential mode"), simply by supplying all the water and N the crop is calculated to need. The other is run using actual soil conditions (Figure 8.). The

Figure 7. The Potato Calculator front page.

difference in N uptake is taken as the first approximation of the seasonal fertilizer N required. A set of user-defined rules is then applied to decide on the maximum size of the N-fertilizer application (Figure 9).

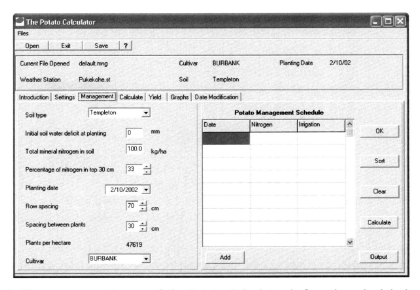

Figure 8. The management page of the Potato Calculator before the schedule has been calculated.

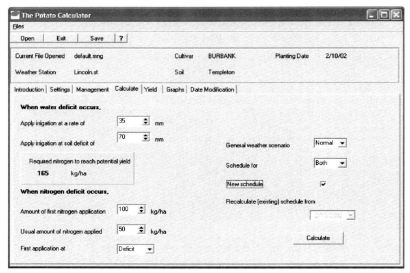

Figure 9. The Calculate page of the Potato Calculator showing user-defined settings and calculated required N.

The resulting schedule (Figure 10) raises the predicted yield to reach its potential. A graphics page (Figure 11) enables the user to review the effects on performance of the

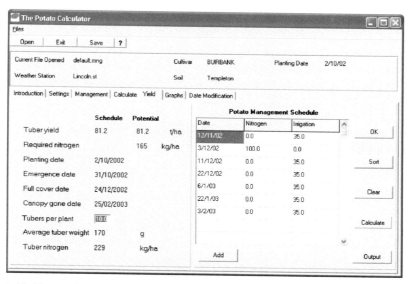

Figure 10. Yield page from the Potato Calculator showing fertilizer and irrigation schedule. The schedule increased the predicted yield from 49.5 to 81.2 t ha^{-1}.

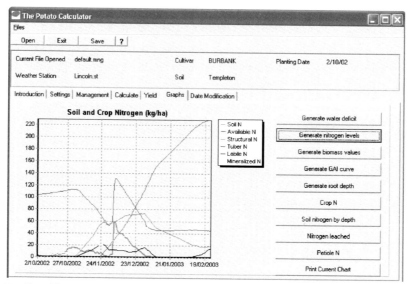

Figure 11. Graphics page of the Potato Calculator showing soil and tissue N content predictions.

management changes made. The schedule (Figure 10) is editable and interactive, and will keep a record of management to date, only rescheduling future options in the light of changes in the weather.

Data requirements

The Calculator needs daily weather records of solar radiation, maximum and minimum temperatures and precipitation as a minimum. If daily wind run and humidity data are available, it will use them in the calculation of Penman evapotranspiration (French and Legg, 1979). If they are absent then it uses the simpler Priestley and Taylor (1972) model.

The soil description includes soil hydraulic properties by depth, soil organic N content and a mineralization coefficient. Initial conditions require an assessment of soil mineral N and its distribution in the upper metre of soil.

Implementation

During October and November, 2002, the Potato Calculator was installed on the computers of five major potato growers who supply Russet Burbank potatoes for processing. Part of a potato crop on each farm was isolated to test the Calculator's predictions. A randomised complete block experiment (three replicates) was set up with three treatments - no added N, calculator-recommended N and normal practice. On these farms normal practice is to supply approximately 300 kg N ha^{-1} as 100 kg N ha^{-1} at planting and the rest in 50 kg N ha^{-1} increments when indicated by petiole nitrate tests (Kleinkopf *et al.*, 1984). It is not their practice to sample the soils before planting. However, as part of this exercise, we sampled mineral N down to 80 cm depth, or to stones where these occurred at a shallower depth (Table 2). In most cases soil mineral N supply was substantial, and the variation between the lowest and highest supply was five-fold. In those cases where there was a substantial mineral N supply, a good proportion (43-82%) was below 20 cm.

Table 2. Soil mineral N (kg ha^{-1}) by depth at the experiment sites.

Site	Soil type	Depth (cm)				
		0-20	20-40	40-60	60-80	Total
Pye	Chertsey silt loam	76	60	20	-	156
Hewson	Chertsey silt loam	48	35	50	-	133
Scott	Waimakariri silt loam	22	10	2	2	36
Newton	Templeton silt loam	64	22	12	14	112
Howie	Templeton silt loam	31	53	47	43	174

All of the growers applied 100 kg N ha^{-1} at planting (including, because of a communication breakdown, in the "nil" plots). The Calculator recommended that Hewson, Newton and Howie should apply 50 kg N ha^{-1} during late January or early February, that Scott should apply two increments of 50 kg N ha^{-1} in mid-January and early February, and that Pye should apply no more N. These are totals of 100-200 kg N ha^{-1}, substantially less than "normal practice". Given that Martin *et al.* (2001a) showed that some treatments whose petiole N tests indicated inadequate fertilization yielded no less than those treatments whose tests were adequate and levels of fertilization were higher, it is likely that the "normal practice" treatments will get substantially more N than the calculator recommendations. At the time of writing (early January 2003), two growers have already exceeded Calculator recommendations for total N application by 60 kg N ha^{-1}. Because the experiments are still underway, we cannot report the results.

Afterword

One of the reasons this project has been enthusiastically supported by both industry and government is because it provides a method of demonstrating good stewardship of land and groundwater. Like farmers in many countries, New Zealand growers would like maximum freedom to decide their own crop management, and most are concerned citizens who do not want to cause environmental problems. The Potato calculator and the related Sirius Wheat Calculator are able to predict the state of the soil and the risk of leaching either during or after the cropping phase (Figure 6). N fertilizer and irrigation can be managed to minimise leaching risks without compromising profit. Our collaborating growers hope to forestall regulation of the absolute amount of N they are allowed to apply by being able to demonstrate good practice that will avoid significant risk to groundwater.

The Potato Calculator is yet in its infancy. The first test is of its credibility. Our experience with the Sirius Wheat Calculator (Jamieson *et al.*, 2003) has been fortunate - either very close correspondence between measured and predicted yield, or clear reasons for discrepancies. We have asked our collaborating growers to be tough with us - to "break" the Calculator if they can. We have also asked them to evaluate the software critically - to suggest the things that they want rather than the things we think they need. Once again, our experience in the first year of release of the Sirius Wheat Calculator led to substantial changes in the way we presented information, and in some changes to the type of information. Some of these lessons have been applied to the Potato Calculator. However, we are certain we have more to learn.

Acknowledgements

Funding for the development and implementation of the Potato Calculator was from several sources. The New Zealand Foundation for Research, Science and Technology funded the underpinning science. The New Zealand Vegetable and Potato Growers

Federation, McCain Foods, Ballance Agri-nutrients and the New Zealand Ministry of Agriculture and Forestry Sustainable Farming Fund provided support for the implementation and on-farm testing of the calculator. We are grateful for the cooperation and support of the collaborating growers, Ross Hewson, Tony Howie, Alan Newton, Dean Pye and Peter Scott.

References

Allison, M.F., E.J. Allen & J.H. Fowler, 1999. The nutrition of the potato crop. British Potato Council Research Report 807/182. 92 pp. (Available at Http:/www.potato.org.uk)

Biemond, H. & J. Vos, 1992. Effects of nitrogen on the development and growth of the potato plant. 2. The partitioning of dry matter, nitrogen and nitrate. Ann. Bot., 70, 37-45.

French, B.K. & B.J. Legg, 1979. Rothamsted irrigation 1964-1976. J. Agric. Sci., Camb. 92, 15-37.

Gardner, B.R. & B.R. Jones, 1975. Petiole analysis and the nitrogen fertilization of Russet Burbank potatoes. Am. Potato J., 52, 195-200.

Grindley, D.J.C., 1997. Towards an explanation of crop nitrogen demand based on the optimization of leaf nitrogen per unit leaf area. J. Agric. Sci. Camb. 128, 377-396.

Jamieson, P.D., 1985a. Irrigation response of potatoes. In: Potato growing: a changing scene, G.D. Hill and G. Wratt (Eds.), Agron. Soc. N.Z. Special Publication No 3, 17-20.

Jamieson, P.D. 1985b. Soil moisture extraction patterns from irrigated and dryland crops in Canterbury. Proc. Agron. Soc. N.Z., 15: 1-6.

Jamieson, P.D. & R. Genet, 1985. Potatoes (table). Planting and cultivation for Canterbury conditions. Aglink FPP 109. 2 pp.

Jamieson, P.D., 1999. Drought effects on transpiration, growth and yield in crops. J. Crop Production, 2, 71-83.

Jamieson, P.D., M.A. Semenov, I.R. Brooking & G.S. Francis, 1998. Sirius: a mechanistic model of wheat response to environmental variation. Eur. J. Agron., 8, 161-179.

Jamieson, P.D. & M.A. Semenov, 2000. Modelling nitrogen uptake and redistribution in wheat. Field Crops Res.68, 21-29.

Jamieson, P.D., T. Armour & R. Zyskowski, 2003. On-farm testing of the Sirius Wheat Calculator for N fertilizer and irrigation management. Proceedings 11th Australian Agronomy Conference, Geelong, February 2003. (In press).

Kleinkopf, G.E., G.D. Kleinschmidt & D.T. Westermann, 1984. Tissue analysis: a guide to nitrogen fertilization for Russet Burbank potatoes. University of Idaho College of Agriculture Current Information Series No. 743. 3 p.

MacKerron, D.K.L. & P.D. Waister, 1985. A simple model of potato growth and yield. Part 1. Model development and sensitivity analysis. Agric. For. Meteorol., 34, 241-252.

MacKerron, D.K.L., M.W. Young & H.V. Davies, 1993. A method to optimise N-application in relation to soil supply of N and yield of potatoes. Plant & Soil 154, 139-144.

MacKerron, D.K.L., M.W. Young & H.V. Davies, 1995. A critical assessment of petiole sap analysis in optimising the nitrogen nutrition of a potato crop. Plant & Soil 172, 247-260.

Martin, R.J., P.D. Jamieson, D.R. Wilson & G.S. Francis, 1992. Effects of soil moisture deficits on yield and quality of `Russet Burbank' potatoes. N.Z.J. Crop Hort. Sci., 20:1-9.

Martin, R.J., M.D. Craighead, P.D. Jamieson & S.M. Sinton, 2001a. Methods of estimating the amount of N required by a potato crop. Agronomy NZ, 31, 81-86.

Martin, R.J., M.D. Craighead, P.H. Williams & C.S. Tregurtha, 2001b. Effect of fertilizer rate and type on the yield and nitrogen balance of a Pukekohe potato crop. Agronomy NZ, 31, 71-80.

Priestley, C.H.B. & R.J. Taylor, 1972. On the assessment of surface heat flux and evaporation using large scale parameters. Monthly Weather Rev., 100, 81-92.

Ritchie J.T., 1972. Model for predicting evaporation from a row crop during incomplete cover. Water Resources Res. 8, 1204-1213.

Stone, P.J., P.D. Jamieson, I. Sorensen & B. Rogers, 2003. Yield consequences of synchrony or otherwise of potato crop peak green area index and solar radiation. Paper presented at the Third International Potato Modelling Conference, Dundee, March 2003.

Vos, J. & P.E.L. van der Putten, 1998. Effect of nitrogen supply on leaf growth, leaf nitrogen economy and photosynthetic capacity in potato. Field Crops Res. 59, 63-72

Williams C.J.M. & N.A. Maier, 1990. Determination of the nitrogen status of irrigated potato crops I. Critical nutrient ranges for nitrate-nitrogen in petioles. J. Plant Nutr., 13, 971-984.

Name of the DSS

On-Farm Assessment of Tuber Size Distribution

Who owns the DSS, Principal author's, address

Scottish Crop Research Institute,

Principal Author: Bruce Marshall Invergowrie, Dundee DD2 5DA, Scotland

What is it for, what questions does it address?

- Estimating tuber size distribution
- Forecasting the development of tuber size distribution and thereby determining the optimum time for destroying the haulm

What input is required?

- Tuber samples to be dug
- The length of the row dug
- The average row spacing

Who is it for, who are the intended users?

Agronomist, Growers and Consultants, anyone who observes tuber size distributions in the crop or wishes to determine optimum time for haulm destruction based on tuber size distribution

What are the advantages over conventional methods (whatever those may be)?

- More accurate, precise and quicker measurements of tuber size distribution in the field without the need for detailed grading facilities
- Estimates critical parameters needed to forecast the development of tuber size distribution and adds value to the data collected

How frequently should the DSS be used or its values be updated?

Depends how critical the crop is and how frequently the user normally samples his crop for assessing size distribution. Could be 3 to 5 times during the main tuber bulking phase

What are the major limitations?

- No major limitations
- Requires a digital camera which could have other purposes for the business
- Samples may have to be brought back to an area shaded from direct sunlight

What scientific / technical enhancements would be desirable?

- Ascertain the optimum and most practical method for taking the digital images on the farm and, in tandem, develop the image analysis software to cope with less than optimal image with quick and easy to use interface for the grower to modify the image if necessary
- Interface the resultant data to a tuber size distribution forecasting system such as MAPP or cut down version

7. Automated on-farm assessment of tuber size distribution

Bruce Marshall and Mark W. Young

Failure to meet precise size specifications has been estimated to cost the UK potato industry £24 million per annum. Accurate and detailed assessment of tuber size distribution during the growing season is difficult and time consuming but of great value to the industry. We present a solution to this problem - digital cameras can provide images, on farm, of samples of tubers dug from growing crops. From these images, with the aid of suitable computer software, individual tuber sizes (mm) and weights (g) and hence the number and total yield of daughter tubers greater than 15 mm are estimated. Computer aided analysis of the resultant images provide measurements of area, length, width and perimeter of each tuber. The best predictor of individual tuber weight is the area and that of depth is the width of the recorded tuber images. A new and more precise method of estimating riddle size from the recorded tuber images is based on both the area and the length:width ratio. This new method does not require any prior knowledge of the cultivar. Its resolution approaches that of the direct method of physical grading and is approximately double that achieved using tuber weight alone as a predictor - a method currently used in the industry. The science is proven. All that remains is to develop a more user friendly interface for the image detection and provide a practical guide to its use on the farm e.g. using the best backgrounds and lighting conditions to produce images that require less or no intervention by the user during its subsequent analysis.

Background

The specialization of potato markets has increased the demand for specific tuber sizes. As potatoes are grown for seed and at least one other market, e.g. baker, pre-pack, crisping, punnet etc, there are at least two and usually more sets of tuber size requirements for each cultivar. Failure to achieve the correct size range of tubers for the specified market is the single most important cause of economic loss in the potato industry. The cost to the UK potato industry of not meeting size specifications has been estimated at £24 million per annum (British Potato Council, 1999). Accurate and detailed assessment of tuber size distribution during the growing season is difficult and time consuming. Marshall (2000) reported on an improved predictive model of tuber size distribution, based on the gamma distribution, that predicts both yields and numbers of tubers in user-specified grades. This model outperforms previous published models (Marshall *et al.*, 1992; Nemecek *et al.*, 1996; Sands and Regel, 1983) and for the first time predicts numbers of tubers in a grade as well as their total weights. The model also confirms that the distribution of individual tuber sizes can be predicted accurately knowing only the number and total weight of tubers greater than 15 mm. While some cultivars were significantly more uniform in their distribution than others, the differences were likely to be small compared to local variations in conditions within a field. The possibility of using directly observed distributions, rather than any formal mathematical function such as the gamma distribution, was also considered. The last method also requires a measure of size distribution. An easy and robust means of obtaining this information is required.

The appearance of digital cameras with megapixel image size (approx. 1000 x 1000 pixels) at prices which are now around €500 (and falling) offers the opportunity to capture valuable information and improve the precision with which tuber size distribution can be measured and predicted. Potato growers often take repeated digs of tuber samples as the crop is thought to be approaching the optimum size distribution. This provides a golden opportunity to record these samples digitally, and extract more detailed and precise information as to the shape, number and distribution of tuber sizes present. Commercial software such as MAPP (MacKerron *et al.*, this book) can then make better use of this information to forecast the best time for haulm destruction in order to achieve the optimum size distribution for the intended market. We envisage that the grower will be able to photograph the dug tubers either *in-situ* against the disturbed soil surface or, at worst, bring them into a barn to photograph them out of strong sunlight. Strong shadows created against some backgrounds in bright sunshine can make subsequent image analysis difficult. The images are then transferred to the farmers PC, analysed, and the data stored in a suitable form for importing directly into a decision support system for managing tuber size distribution. As a bonus, the grower also has a permanent record of the individual tubers sampled for future reference and can also use the camera for other purposes such as incidence of foliar disease, recording ground cover etc. But first, it is important to get the science right. The research reported here proves the method is feasible and practical. It identifies the best methods of estimating individual tuber weight (g, fresh weight) and grade size (to the nearest 5 mm square riddle) to which each tuber belongs. Given these methods, it is then possible to determine the total weight of tubers greater than 15 mm, the total number of tubers greater than 15 mm and the grade size distribution directly from digital images of samples dug on the farm. Finally, the practicalities of obtaining suitable images on farm are assessed using a range of backgrounds and lighting conditions.

System development

Tuber material

Potatoes were collected from experimental material at Scottish Crop Research Institute (SCRI) and various external sources, produced in 1999 (Table 1). Material from a total of 14 cultivars was collected and covered the range of shapes from 3 through to 8 on

Table 1. List of cultivars with their National Institute of Agricultural Botany (NIAB) score for shape, showing the total number of 5mm-wide grades, the minimum (Min) and maximum (Max) size of tuber and the total number (n) of individual tubers studied by cultivar and source. The dimension mm refers to the square riddle size.

Cultivar	Source	NIAB shape	Number of grades	Min (mm)	Max (mm)	n
Cara	SCRI	3	12	15	75	211
Charlotte*	SCRI	-	8	15	50	153
Pentland Dell	Greenvale	7	7	20	50	140
Desiree	SCRI	7	14	15	85	235
King Edward	SCRI	6	14	15	85	228
Estima	SCRI	6	11	15	70	205
Marfona	Taypack	4	14	20	90	242
Nadine	Newlandhead	4	15	15	90	261
Maris Peer	SCRI	4	10	15	65	163
Maris Piper	Greenvale	4	9	20	65	164
Maris Piper	SCRI	4	12	15	75	210
Romano	Taypack	3	10	30	80	200
Saturna	Greenvale	3	7	25	60	140
Shepody	SCRI	8	13	15	80	213

* Charlotte is not listed in the NIAB tables.

the National Institute of Agricultural Botany (NIAB) listings. The NIAB score for shape ranges from 0 (round) to 9 (long oval) and the hundred-plus cultivars currently on the list all fall into the range 2 through 9. Tubers from each cultivar were sized by hand through square riddles from size grade 15-20 mm upwards in 5 mm increments to 85-90 mm. Each sample of cultivar x size grade combination (up to a maximum of 20 tubers per sample) was stored in separate, labelled bags at 4°C.

Image capture, processing and measurement

Each sample was taken out of store, placed on to a labelled template of matt black material, the size grade and cultivar were marked on the template (Figure 1) and then the sample was photographed. The template has positions for up to 20 tubers within a 75 cm square marked by its corners. The tubers were carefully placed with their shortest axis being vertical i.e. lying on their flattest side. The camera, a Kodak Digital Science DC210 Plus, was held in a tripod and set with a spirit level to look vertically down onto the template, thus minimising aspect errors. For convenience, the height of the camera above the template was kept constant from one day to the next to maintain the same image resolution. The photographs were taken indoors with the aid of the in-built flash.

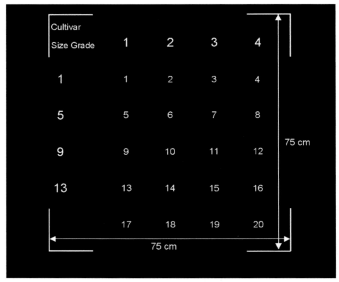

Figure 1. Template for the placement and photographing of up to 20 tubers of known individual weight and belonging to a specified, 5 mm grade size and cultivar.

Image capture was set to JPG format to minimise file size. After each image was taken, the weight and vertical height of each tuber in the image was then recorded.

An image acquisition and analysis package, Scion Image (Release: Beta 4.0.1), was used to analyse the resultant images. Scion Image is available from http://www.scioncorp.com. The digital images were converted from colour, JPG to grey-scale, TIF images for speed of image processing.

The number of pixels between each marked corner on the template was noted for each image. The mean of these values was taken and used to calibrate the measurements made by Scion Image. The calibration was typically 1 mm per pixel and was invariant from one image to the next to less than 1 per cent.

The image processing steps were as follows
- Filtering to reduce noise on the edges of tubers (a simple smoothing algorithm)
- 'Thresholding' to detect the tubers from the background (a single, grey-scale threshold value)
- Conversion of the detected image to a binary format (tubers appear white on black background)
- Inversion of the "look up table" to produce a negative image (tubers then appear as black objects on a white background)
- 'Particle analysis' of the black objects - the automatic selection and measurement of the individual particles,
- Output of results to a file for further analysis (Table 2).

Table 2. For each combination of cultivar and size grade image, the following parameters were recorded in the Scion Image output file for each tuber. The Designation indicates the term used in the text to refer to these measurements.

Scion image output	Designation	Notes
Identification number		Particle number
Area	area	True area of each identified particle
X-Y centre		X and Y co-ordinate of the centre of each particle, the origin is the top, left-hand corner of the image
Perimeter/length	perimeter	True perimeter length of each particle
Ellipse major axis	length	Major axis of the best ellipse whose area is equal to the true area of the particle
Ellipse minor axis	width	Minor axis of the best ellipse whose area is equal to the true area of the particle

A macro was written that draws the major and minor axes of the best fitting ellipse to each detected particle (Figure 2). In most cases the lengths of major and minor elliptical axes of the particle are very close to the true maximum length and the corresponding maximum width at right angles to that length. Occasionally, a short length of stolon can remain attached to the base of a tuber. This causes the true maximum length of the tuber to be overestimated. For this reason the major and minor axes of the best fitting ellipse were used as more robust measures of tuber length and width respectively. The

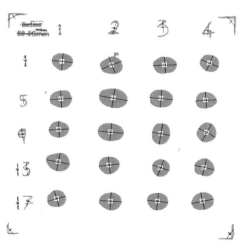

Figure 2. The detected image with particles greyed out and identification numbers overwritten. The major (designated length) and minor (designated width) axes of the best fitting ellipses are overdrawn on the detected particles.

chalked numbers and marks are also detected as "particles" at this stage. These particles were deleted later, by hand, from the output files. However, the removal of such "particles" is a relatively straightforward task to automate by the inclusion of a minimum area requirement and setting tolerances on the shape of the object.

The Scion Image output files were then read into Excel and the data for each tuber extracted. In the analyses, the Scion Image results are referred to by their designated terms listed in Table 2. The metric perimeter was recorded and although strongly correlated with weight was not as good as other predictors, especially area, and is omitted from the paper for clarity. After each digital image was captured, the vertical height (designated as depth) and the fresh weight (designated weight) of each tuber along with its location number on the template were recorded.

Evaluation

Data overview

In total, 2765 individual tubers were photographed and their fresh weights and depths recorded (Table 1). A logarithmic transformation was applied to all the data for two reasons: (1) the variation about a correlation between any pair of metrics was found to increase with their mean values, and (2) the relations between specific pairs of metrics using the raw data were non-linear. The dependence of variability on mean value is a common occurrence in biological properties that relate to size and is removed by the logarithmic transformation. Non-linearity is also to be expected for specific pairs of metrics. Length, width and depth are all one-dimensional scales of length, whereas area is two-dimensional and weight, which is proportional to volume, is a three-dimensional scale of length. Thus relations between any pair from length, width and depth will be expected to be linear, but are non-linear when any one of these linear parameters is compared with area (an approximately quadratic relation is expected) or weight (an approximately cubic relation is expected). Logarithmic transformation linearises these relations, as well as achieving constant variance. All the Scion Image parameters and those observed directly on the tuber are highly correlated, irrespective of cultivar. A typical correlation matrix for cultivar Pentland Dell is shown in Table 3. The correlations increased when moving from untransformed to transformed data in every cultivar.

The strong correlations in Table 3 indicate that the Scion Image parameters are all individually, good predictors of tuber weight - the best predictor being area, as judged by the largest correlation of 0.989. Likewise width is the best single predictor of depth.

Predictors of tuber depth

As indicated in Table 3, the best single predictor of depth is width. For each cultivar separately, linear regressions of $\log_{10}$(depth) on $\log_{10}$(width) accounted for 86.9 to 97.3 per cent of the variance (average 94.4%, p < 0.001 in all cases). Linear regressions of $\log_{10}$(depth) on $\log_{10}$(length) accounted for less of the variance, 81.0 to 94.6 per cent

Table 3. Correlation matrix of the log-transformed values of the designated parameters. The data is for the cultivar Pentland Dell. The first three parameters, area, length and width, are estimated by the Scion Image software from the stored digital images. The last two parameters, depth and weight, are measured directly from the tubers.

		Image			Tuber	
		Area	Length	Width	Depth	Weight
Image	Area	1				
	Length	0.987	1			
	Width	0.979	0.960	1		
Tuber	Depth	0.947	0.934	0.970	1	
	Weight	0.989	0.963	0.967	0.952	1

(average 89.1%). Interestingly, $\log_{10}$(area) was almost as precise a predictor as $\log_{10}$(width), accounting for 85.9 to 96.7 per cent of the variance (average 93.2%). These results suggest that there is little variation across the cultivars studied in the linear relations between $\log_{10}$(depth) on $\log_{10}$(width). Figure 3 confirms this. The individual estimates of slope and intercept for each cultivar lie close to a common relation and the shape of the cultivar, as indicated by the NIAB score, has no systematic influence on these estimates. The common trend-line accounted for 96 per cent of the variance in the parameter estimates.

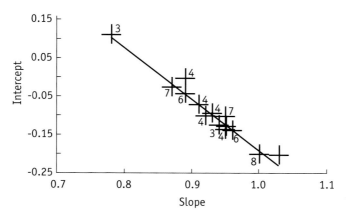

Figure 3. The relation between the two parameters, intercept and slope, of the linear regression of $\log_{10}$(depth) on $\log_{10}$ (width) estimate for each cultivar separately. The numbers indicate the NIAB score for "shape" and the crosses indicate the average standard errors. A linear trend line has been added. The unlabelled cross is for the cultivar Charlotte.

It is common for the estimates of slope and intercept to be strongly correlated with much of the variation in their estimates being due to random sampling rather than true differences between cultivars. In other words, if one assembles a large collection of data for a single cultivar and takes repeated, random sub-samples of data of equal size from the collection, then the estimates of slope and intercept will be strongly correlated. This effect is embedded in the results shown in Figure 3 along with any real differences between cultivars. A more detailed investigation of the individual regression lines, $\log_{10}$(depth) on $\log_{10}$(width), revealed that the average ratio of depth:width was 0.615 with a coefficient of variation of only 5.9 per cent across the cultivars concerned. Pooling the data across all cultivars showed that a common relation accounted for 93.1 per cent of the variation and that while allowing variation in slopes and intercepts among cultivars was significant ($p < 0.001$) the improvement was small, accounting for only a further one per cent of the total variation in $\log_{10}$(depth).

If the depth:width ratio remains constant as tubers grow then the expected value of the slope is unity. A value less than unity indicates a flattening of the tuber as it expands. The value of the slope averaged over all cultivars was 0.93, which indicates a reduction of 12 per cent in the depth:width ratio as a tuber grows from 15 to 85 mm. In two cultivars, Charlotte and Shepody, the slope was not significantly different from unity. With Charlotte it must be borne in mind that the range of tuber sizes available was limited, however the size range for Shepody was fully represented (Table 1).

Examination of the corresponding relations between estimates of intercept and slope for $\log_{10}$(length) and $\log_{10}$(area) (not shown) revealed systematic deviations from a common, linear trend-line which was related to the NIAB score for shape. Cultivars with low scores tended to have intercept values that fell above the common line and vice versa. These results are consistent with differences in shape between the cultivars studied being manifest primarily through changes in length relative to depth or width, rather than in changes in depth:width ratio.

Predictors of tuber weight

As indicated in Table 3, the best single predictor of weight is area. Taking each cultivar separately, linear regressions of $\log_{10}$(weight) on $\log_{10}$(area) accounted for 98.0 to 99.6 per cent of the variance (average 99.2%, $p < 0.001$ in all cases). Linear regressions of $\log_{10}$(weight) on $\log_{10}$(length) still accounted for 95.6 to 98.5 per cent of the variance (average 96.8%). Surprisingly perhaps, but consistent with the correlations in Table 3, $\log_{10}$(width) was a more precise predictor than $\log_{10}$(length), accounting for 97.1 to 99.2 per cent of the variance(average 98.2%).

Adopting a similar analysis as for tuber depth produced a similar relation between the estimates of slope and intercept of the linear regression of $\log_{10}$(weight) on $\log_{10}$ (area) (Figure 4). Again, much of the variation in estimated values was due to the "sampling effect". Pooling the data across cultivars showed that a common relation accounted for 99.2 per cent of the variation while including slopes and intercepts among cultivars in the analysis accounted for only an additional 0.1 per cent of the total variation in $\log_{10}$(weight). Differences between cultivars were clearly very small but nevertheless they were statistically highly significant (p < 0.001) and the effect of cultivar shape on the relation is evident. Cultivars with a low NIAB score (rounded tubers) tending to fall above the common relation and those with a higher NIAB score (increasingly long oval) tending to fall below the line (Figure 4). In this case, if the depth:width ratio remains constant as tubers grow then the expected value of the slope is 1.5 or 3/2 - volume (cubic) / area(square). A value less than 1.5 is consistent with either a reduction in depth:width ratio(flattening) or an increase in length:width ratio or both, as the tuber grows. Since at least some of the cultivars did not exhibit flattening in the earlier analysis of depth and none of the cultivars achieved a slope value of 1.5 here, then at least some of the cultivars exhibited changes in the length:width ratio as their tuber sizes increased.

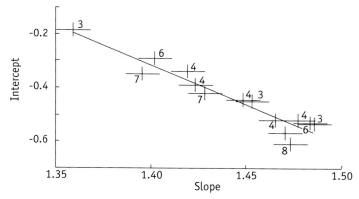

Figure 4. The relation between the two parameters, intercept and slope, of the linear regression of log $_{10}$(weight) on log$_{10}$ (area) estimate for each cultivar separately. The numbers indicate the NIAB score for "shape" and the crosses indicate the average standard errors. A linear trendline has been added. The unlabelled cross is for the cultivar Charlotte.

Estimating riddle size

The conversion between riddle size (mm) and tuber fresh weight (g) or its surrogate, area, can be done by estimating the linear relation between their logarithmically transformed values or by fitting a power function directly to the untransformed data. The distribution of tuber weights, and likewise area, found in any one 5 mm size range has considerable overlaps with neighbouring size ranges. Figure 5 shows an example for the cultivar Cara. The same value of area frequently occurs in three and even four adjacent grades. The reason for this is that variations in tuber shape, particularly in length:width ratio, result in a wider range of tuber weights (and areas) falling into the one grade than would occur if shape was invariant. The tubers were carefully graded by hand for this experiment to avoid the grade "jumping" that occurs in mechanical grading

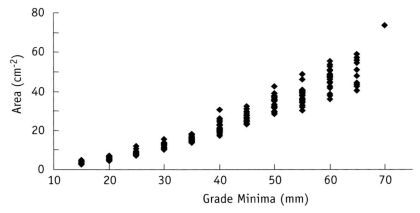

Figure 5. The relation between area, estimated from Scion Image analysis of individual tubers and the size range from which they came. The size ranges are 5 mm wide and the minimum square riddle size of each grade is indicated.

i.e. tubers failing to be stood up on their long axis and hence passing over a riddle size they should have normally fallen through.

The relation between riddle size and weight or area also changes with cultivar because of their known differences in shape (Figure 6). Marfona has a NIAB shape score of 4, which is more rounded than the longer, oval shape of Shepody that has a NIAB shape score of 8. Individual conversion functions are therefore required for at least each class of cultivar grouped by their NIAB shape score (scale 0 to 9). However, such conversion functions cannot take into account variations in shape between individual tubers from the same cultivar and size grade.

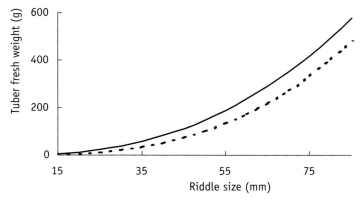

Figure 6. Relation between tuber weight and riddle size for Shepody (solid line) and Marfona (dashed line). The lines are the back transformed values from linear regressions fitted to the logarithmic values of tuber fresh weight and mean riddle size.

Ideally, a method is required that can classify tubers into the appropriate riddle size using only the information available in the image, without necessarily requiring knowledge of the cultivar. Canonical Variance Analysis (CANVA) was used to identify potential discriminators between cultivars. The three Scion Image parameters, length, width and area, were used to create a three dimensional space in which each tuber is represented as a point corresponding to that individual's parameter values. In CANVA, the space is rotated to the direction that maximises the variation between groups compared to the variation within groups. This direction becomes the first canonical variate. The process is then repeated with the remaining variation. In general, with n parameters there are n -1 canonical variates. In this case, with three parameters, there were just two canonical variates.

The results of CANVA speak for themselves (Figure 7). The analysis was set up to discriminate between cultivars. Not only are the cultivars discriminated by the first canonical variate, but there is also a systematic trend from left to right of increasing score for NIAB shape (Figure 7a). Several cultivars have similar shapes and tuber shape also varies within cultivars. Not surprisingly, there is considerable overlap of data between cultivars and for clarity, only the cultivar means and their standard errors are shown in this first view. In the second view (Figure 7b), the same data is re-presented as the individual data points and re-labelled according to the riddle size into which each tuber fell. It is important to realise that this is the same CANVA as was set up to discriminate between cultivars and not between riddle sizes. In fact, it has achieved both discriminations. In this second view, one can clearly see a systematic separation of tubers into riddle sizes in the direction of the second canonical variate, CV2. Furthermore, in contrast to the earlier classification, based on weight or area alone (Figures 5 and 6), most tubers can now be accurately classified into a single, 5 mm grade and all tubers into at least one of two grades. This represents an increase in resolution of more than one whole grade over the previous method.

Examination of the loadings of three parameters that comprise both canonical variates reveals that the first axis, CV1, is equally dominated by the logarithms of length and width and the negative sign for width indicates that the score for CV1 is determined by the difference, $\log_{10}$(length) - $\log_{10}$(width) (Table 4). This is equivalent to $\log_{10}$(length/width) i.e. it is a measure of the shape and consistent with the left to right trend in NIAB shape factor observed in Figure 7a. A little care is required in interpreting loadings generated by CANVA. The scales upon which the individual parameters are measured influence the magnitudes of the loadings. In this case, both length and width are measured on the one scale and so the magnitudes of their loadings can be compared directly. In the case of area, the same scale of length is used but in two dimensions i.e. length squared. Since the data is first transformed to the logarithm, the variance in the area parameter will be approximately twice as large as that for either length or width. Consequently, the same loading for area as that for the other two will have twice as much impact. Even taking this into account, CV1 is still dominated by $\log_{10}$(length/width). In CV2 the magnitude of the loading for area is about half that for length and more than three-times that for width. Taking into account the factor of two for area and the fact that the sign of the loadings for both area and length are negative

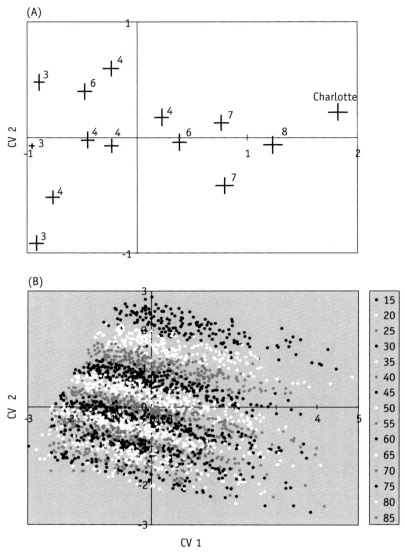

Figure 7. Canonnical variance analysis of the logarithms of length, width and area grouped by cultivar (a) the mean and standard error of scores for all tubers belonging to a specified cultivar along with corresponding NIAB shape score for that cultivar and (b) the same data re-presented as individual tubers and re-labelled with the 5 mm riddle size to which the indivdual tuber belongs. CV1 and CV2 are first and second canonical variates.

Table 4. Loadings of the three Scion Image parameters used in the canonical variance analysis of individual tubers grouped by cultivars. CV1 and CV2 are the first and second canonical variates.

	CV1	CV2
$Log_{10}(area)$	5.4	-18.3
$Log_{10}(length)$	193.2	-39.4
$Log_{10}(width)$	-204.7	5.5

then CV2 is proportional to the sum, log_{10}(area) + log_{10}(length), which is equivalent to log_{10}(area*length). Area*length has the dimensions of length cubed. Thus CV2 is the logarithmic transform of a volume metric and, assuming that variation in tuber density is small, could be said to reflect a measure of tuber weight. It is therefore not surprising, perhaps, that it is a useful discriminator of riddle size. The negative sign of their loadings indicates why data for the smallest grades is at the top of the graph and the largest is at the bottom (Figure 7b). Strictly, because the distribution of data for any one grade is not absolutely parallel to CV1 (it slopes slightly down from left to right), CV2 on its own is little better or no better than either weight or length. It is only when the value of CV1 is also taken into account that the improvement in the resolution of grade size is achieved.

Factors influencing detection of tubers - Backgrounds and digital filters

Tubers were photographed on the various backgrounds (black, blue, red, green, yellow indoors and black polythene, concrete, dry soil, grass and wet soil outdoors). The resultant colour images were loaded into the Scion Image software and then separated into their three component colours RGB (Red, Green and Blue) and also directly converted to a grey-scale image. In Scion Image, a cross-section was drawn across the middle of the largest tuber on each RGB and grey-scale image so that a portion of the background at either side of the tuber was included. The pixel values along the cross-section were then stored.

In most cases the greatest contrast between the tuber and the background was found using the red filter and the worst contrast when using the blue filter. The poorest backgrounds were the green and red colours and the concrete, where none of the filters or the grey-scale provided a good discrimination of the tuber. The performance of the green filter was generally intermediate between red and blue. The best backgrounds were black and blue colours and moist soil, especially with the red filter (Figure 8). Grey-scale images also performed adequately with these backgrounds, similar to that with the green filter. Strongly reflective backgrounds (e.g. black polythene) do not work well, especially if the flash is activated or if there is a strong directional light source.

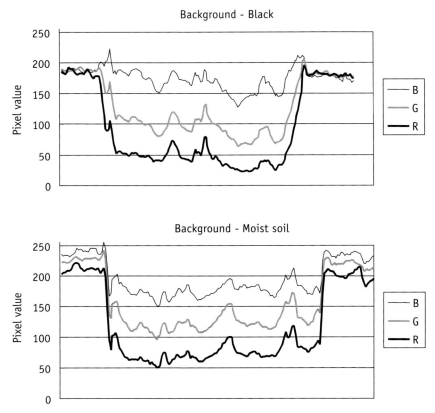

Figure 8. Pixel value (0 ... 255) across a line transect of a tuber photographed against a black background and a moist soil. There is a single trace for each of the three colour components, RGB (red, green and blue). The grey-scale response (not shown) was similar to the green filtered trace in both cases.

Factors influencing detection of tubers - Aspect

For the bulk of the work reported here, the camera was carefully set to be parallel to the template with the aid of a spirit level and tripod. Such careful positioning is not practical on the farm. Tilting of the camera results in parallax, where the image of the template, although square in reality, will appear larger on one side than the other. To assess how big a problem this might be, a series of 9 hand-held images of the template were taken with the viewfinder "squared up" by eye using the corners chalked on the template as guides. The aspect errors were no greater in the 9 hand-held images than those recorded with the 154 tripod-based images. The variation in estimated length from one side to the other was no more than 0.9 per cent on average and a maximum of 2.8 per cent was recorded. The sole advantage of the tripod system for experimental purposes was the maintenance of a constant distance between camera and template and hence a constant calibration from one image to the next taken. The calibration was typically around 1 pixel per mm of distance on the template. In any practical implementation, the software will be expected to automate or at least semi-automate this calibration process.

Summary and future development

- Images of dug samples of tubers can be captured using digital cameras with 1 mega-pixel or greater image capacity
- The data is of sufficient accuracy and precision to estimate total number and weight of tubers automatically with the aid of simple image analysis software.
- Given these estimates, both current tuber size distribution and how it will change over time as total yield increases can be accurately predicted.
- In addition, tests have shown that the weight and riddle size of each tuber in the image can be estimated without any prior knowledge of the cultivar.
  - The best predictor of tuber weight was found to be the area of the recorded tuber image
  - The best predictor of riddle size is a combination of the area and the length:width ratio of the recorded tuber image.
  - The precision with which riddle size can be estimated is as good as physical grading and double the resolution achieved using tuber weight alone as predictor - a method currently used in the industry. The latter method also requires a calibration curve for the individual cultivar.
  - The new method is able to provide a detailed description of individual tuber size distributions giving both the number and weights of tubers in each grade, specified either by mm riddle size or weight. This opens up the additional possibility of using the directly observed tuber size distributions to predict future evolution of the distributions up to harvest without imposing a predefined functional form (Marshall, 2000).
  - Such a predictive system could be of considerable benefit in the high-value markets of small -'punnet' and salad potatoes, and in estimating tuber counts in seed crops.
  - The predictive power of current models, with their pre-defined shapes for the tuber size distribution, are at their weakest in these areas.
- There are a two main issues to be addressed before wider release to the industry
  - Can satisfactory images of the tuber samples be captured at the point of lifting or will they have to be brought back to a shaded area e.g. into a farm building, to obtain shadow free images. Assessment of different backgrounds suggests that moist soil, which is frequently present when tubers are dug, and grass (field margin) can provide suitable backgrounds.
  - Develop a graphical user interface with simple, quick editing features to underlying image analysis routines.

Acknowledgements

This work was funded jointly by the British Potato Council and Scottish Executive for the Environment and Rural Development. We thank Laura Marshall for carefully recording the tuber samples along with the digital photographs.

References

British Potato Council, 1999. Research and Development Strategy Document, January 1999.

Marshall, B., 2000. MAFF Final Report for Project Code HP0210T, A predictive model of potato size distribution and procedures to optimise its operation, 1997 - 2000.

Marshall, B., H.T. Holwerda & P.C. Struik, 1992. The influence of temperature on leaf development in potatoes (*Solanum tuberosum*): a statistical model. Field Crops Research 32: 343-357.

Nemecek, T., J.O. Derron, O. Roth & A. Fischlin, 1996. Adaptation of a crop-growth model and its extension to a tuber size function for use in a seed potato forecasting system. Agricultural Systems 52: 419-437.

Sands, P.J. & P.A. Regel, 1983. A model of the development and bulking of potatoes (*Solanum tuberosum* L.). V. A simple model for predicting graded yields. Field Crops Research 6: 25-40.

Name of the DSS

Management Advisory Package for Potato, MAPP

Who owns the DSS, Principal authors, address,...

Owners of IP: British Potato Council & Mylnefield Research Services

Principal Authors: DKL MacKerron & Bruce Marshall, Scottish Crop Research Institute, Invergowrie, Dundee, DD2 5DA, d.mackerron@scri.sari.ac.uk, b.marshall@scri.sari.ac.uk

What is it for, what questions does it address?

Support decisions on :

- Seed rate as modified by cultivar, size of seed and intended market.
- Date of planting.
- Need for irrigation: operational and strategic.
- Manage and interpret results from samples dug from field.
- Time of haulm destruction in the light of developing yield, size grade distribution, and market specification.

What input is required?

MAPP uses the same crop records that a grower would normally keep, together with data on the weather and the specific soil-types in the fields on which potatoes are being grown.

Field data: Soil types and their depths in up to 3 horizons in the soil profile.

Crop data: Cultivar, size of seed tubers, cost of seed, anticipated yield, date of planting, planned date of haulm destruction.

Market data: Size grades specified by intended market(s) and expected, associated prices.

Weather data: • Long term average weather data for the area, at least for the duration of the growing season, or other 'reference' weather data set.
- Also, weather data for the current growing season.
- Irrigation amounts and times.

Who is it for, who are the intended users?

Principally, growers and / or their advisers.

Merchants and Processors wishing to understand the position of their suppliers.

It can also be used (is being used) educationally in a college of agriculture.

What are the advantages over conventional methods (whatever those may be)

It is crop specific. Its guidance is tailored to cultivar and the seed being used and it accommodates multiple market outlets. It does not make single value recommendations but shows the outcome of decisions in a continuum of response. It is initially predictive then accepts progressively more data from the real crop

How frequently should the DSS be used or its values be updated?

Depending upon the application: For most purposes, weekly or twice weekly is adequate. Monthly will not take full advantages of the facilities offered. Minimum is before planting and one month before harvest.

What are the major limitations?

For many growers, obtaining current weather data is the single biggest difficulty.

What scientific / technical enhancements would be desirable?

MAPP cannot use information on fertilizers or soil fertility. Nor does it accommodate effects of pests and diseases.

8. MAPP, the Management Advisory Package for Potatoes: decision support means giving informed options not making decisions

D.K.L. MacKerron, B. Marshall and J.W. McNicol

There are many forms of decision support from published tables of recommendations on, say, fertilizer application or seed-rates for example, all the way to sophisticated computer-mounted systems. However, most examples of all of these do not really offer support. Rather, they offer answers. Essentially, they make the decision for the user and offer little if any scope for the user to ask, "What if I don't do that? What if I make a slightly larger, or smaller, application? What if I know something that you don't?" MAPP, the Management Advisory Package for Potatoes provides an example of a DSS in which the user can examine the consequences of specific judgements such as choice of cultivar, seed size, seed rate, planting date and depth etc. on the crop, given a number of constraints such as the preferred sizes of tuber to be produced and their prices (market specification), soil type, the likely weather, and possibilities for irrigation. The user can then follow the effects of taking actions which do not appear 'ideal' to the system. A feature of the system is that the user can see both the 'optimal' action as judged by the DSS and the consequences of 'sub-optimal' ones. He or she can then apply his or her own attitudes to risk and to money and so elect in an informed manner either to take the 'optimal' action or an apparently 'sub-optimal' one.

Introduction and background

The successful production of potatoes for specific markets involves making many decisions each of which can have consequences for the later growth and yield of the crop, its marketability, and its profitability and the crop and market systems are sufficiently complex that the consequences of a particular judgement are not always evident. Growers appreciate help or support with their decisions but, to be worthwhile, the system offering support must be properly informed.

Knowledge can be divided into broadly two classes, domain and strategic, roughly corresponding to factual knowledge and knowing how to use the domain knowledge. Effectively, simulation models are compilations of strategic knowledge arranged to use a clearly defined set of domain knowledge represented by the parameters and input variables of the model. Single-value outputs from deterministic simulation models, based on closely defined domain knowledge, give a very limited representation of reality, where most of the controlling variables are actually stochastic in nature and the predicted values (such as yield or disease progress) may also be stochastic. There is an implied arrogance in the operation of such systems. The developer may be sure of the domain knowledge in the system, based on experimentation, but that certainty is carried over through the strategic knowledge into another domain, the grower's field in a given year, where it may be inappropriate or inadequate.

Instead, a system that is intended to support a decision-maker should offer the best information that is available and should make evident the best expectations from a range of actions so that the decision-maker can apply his or her own judgement, including factors that may not be appreciated by the designer of the DSS.

Over a period of twenty years, staff at the Scottish Crop Research Institute (SCRI) have developed and applied computer programmes that simulate several aspects of potato growth and that address some of the problems faced by growers (e.g. MacKerron & Waister, 1985; Marshall, 1986, Jefferies and Heilbronn, 1991, Jefferies *et al.*, 1991, Marshall *et al.*, 1995, MacKerron & Lewis, 1995, Rijneveld, 1997). However, the traditional or conventional form of such models has not encouraged their use by farmers. During the 1990s, advances in computer interface systems led to a very wide adoption of computers for many purposes. Concurrently, the technology became available to condense widely distributed knowledge and to present it in a coherent form. In 1997, the project Management Advisory Package for Potato - MAPP was begun at SCRI to draw together differing models into a form that would be understandable by farmers, usable by them, and that could be used to guide their decisions on matters that affect the profitability of their enterprises. MAPP is a single advisory package, mounted in computer-based software, that gives potato growers access to the best knowledge available on certain aspects of the culture of potato and that can assist with their management decisions.

The purposes of this chapter are not only to describe MAPP, but also to emphasise the importance of giving the user the necessary information to make his or her own decisions, based on the output of the models, rather than simply accepting their 'best' solutions. A significant feature towards this is the ability to show the effects of the 'best' solution and of others, in parallel.

Description

The building blocks

The existing models that were available to form a basis for MAPP included: - potential, water-constrained, and forecast yields (MacKerron & Waister, 1985; MacKerron, 1987, MacKerron *et al.*, 1990, Jefferies & Heilbronn, 1991; Jefferies *et al.*, 1991; MacKerron & Lewis, 1995), all of which are time-dependent, weather-driven, semi-mechanistic simulation models. Further, there were models of tuber multiplication and tuber size distribution (TSD) (Marshall, 1986; Marshall *et al.* 1995) which had been developed from the work of Sands & Regel (1983) and combined into an expert system (Smart, 1992), using techniques of artificial intelligence to provide a highly sophisticated version of a seed-rate model, suited to advisors or to skilled management. It was foreseen that the models of forecast yield and tuber size distribution could be combined to provide a model for the prediction of graded yield.

The appearance

MAPP appears to the user through a fully standard Windows interface with standard toolbars and menus of commands and those tools and commands that are unique to the package appear in standard form (Figure 1). The several functions within MAPP are accessed through an 'icon strip' on the left-hand side and also, on some screens, through a 'tab strip' at the bottom of the screen.

The icon strip allows the user to move between choice of the crop to be examined, to examine and to edit the weather data to be used, and to set values of parameters or to ask questions of the system in almost any order that the user may choose. The only restrictions imposed are logical ones provided by the broker (see 'Structure') such that the system will not perform certain functions without the appropriate input data. When and where this occurs, the broker informs the user of the omission.

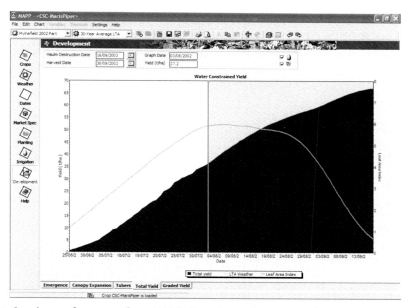

Figure 1. Specimen of a screen from MAPP. Note Status bar names the crop loaded at present, Icon strip selects class of information to be shown, Tab strip selects subset of that class. Menu bar and Tool bar have standard functions plus tools specific to MAPP. Vertical line 'reads' yield and date from graph into boxes near top of screen.

Functionality

The user of MAPP package can receive guidance on:-
- Determining seed-rates taking account of the effects of differing:
  - Cultivars
  - Seed sizes, expressed either in millimetres or in grammes
  - Prices to be paid for the seed
  - Anticipated level of yield
  - Prices to be received for the crop including the use of different prices for different fractions of the crop.
- Irrigation - strategic, tactical, or operational planning
  - Development of soil moisture deficit (SMD)
  - Predicting the future need for irrigations
  - Effects of SMD on canopy development and on yield
- Timeliness of operations:
  - The effects of changing planting date
  - Pre-emergence herbicide application - knowing when emergence will occur
  - The user can also monitor canopy development to check correspondence between the real and the virtual crop.
  - Irrigation, when it is required and the effects on soil moisture deficits and on total and graded yield if deficits are allowed to develop
  - Haulm destruction, when to do it and the effects of choosing another date
- Development of total yield, graded yield, and market value of the crop. So that the user is able to estimate saleable yields and graded yields together with net financial return and optimum date of burn-down of the crop.

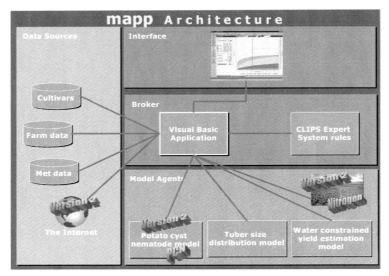

Figure 2. Diagrammatic structure of MAPP.

Design phase

Structure

The two sets of models: the semi-mechanistic (development and yield) and the rule-based (seed rate) are, in a sense, orthogonal and cannot be combined in a single conventional programme. Instead the package MAPP was constructed using a 'Broker and Agent' structure in which a central programme, the broker, interacts, as required, with any from a selection of 'agents'. The agents are models in the conventional sense and they and the broker have access to sources of information in databases (Figure 2).

The development of the system was done under Microsoft Windows 98 and the installation was later checked and implemented for MS Windows 95, Windows NT 4 SP5, Windows 2000, and Windows Me. - It also works in Windows XP. - The principal development tool is MS Visual Basic 6.0, with DAO 3.5 used to talk to a number of MS Access© databases, and a third-party graphing control to provide charting. The TSD model is coded in VB6 as an interfaceless ActiveX control, whereas the Water Constrained Yield Estimation Model is coded in Visual C++ 6.0, also as an interfaceless ActiveX control. There is, also, a small but important rule-base component written in the CLIPS KBS development system.

That rule-base is central to the operation of the whole system as it is what informs the broker on which agents and sources of information are needed for each task. The first stage in developing that rule-base was the construction of decision trees and an influence diagram.

Influence diagrams

The purpose of an Influence diagram is to set out those factors and variables that influence the states of development and the rates of processes. The diagram aims to be mechanistic. That is to say that it should show links between several related items in the order in which the influences operate. It also shows and distinguishes links that are accommodated within MAPP and those that are not treated within MAPP.

Entity queries

An interactive version of the influence diagram provided a useful way to examine relations contained within it but is not included in the standard MAPP package. Double clicking on any box opened up a response from the package to show those factors that influence the feature being examined, both directly and at one remove. It also showed those factors that are influenced by the one of interest. Clicking on the names of any of the influencing factors repeated the exercise for itself. This parallels the influence diagram was used to set up the rule base that governs the operation of the broker within MAPP.

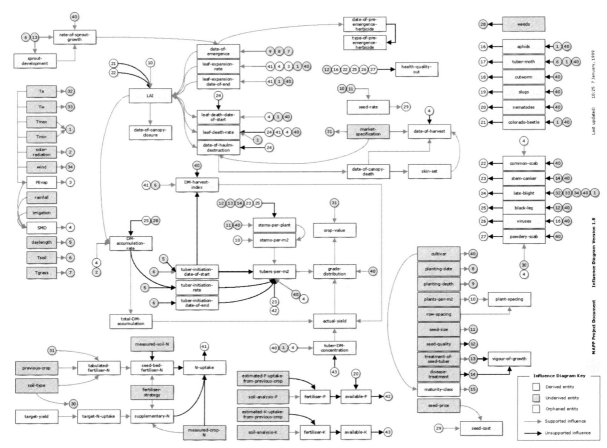

Figure 3. Influence diagram - Diagram of the influences recognized by MAPP and central to its operation.

Primary inputs

A certain minimum of information is required before MAPP will run. This includes crop name, soil data (Figure 4), bed system, row spacing and planting depth, and two sets of weather data - from the current season and, typically, long-term averages.

A second set of data required is called 'Market Spec' or specification. This comprises the several size classes that are of interest to the customer and the prices to be expected for those sizes (Figure 5).

A third set of data specifies the anticipated yield and defines the seed that will be used - Cultivar, size and price of seed and whether chitted or not. Finally, **MAPP** needs to know both when to start and when to stop and so it needs date of planting and date of haulm destruction.

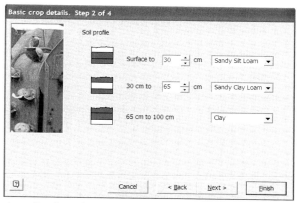

Figure 4. Specification of soil profile by depth and soil type in three horizons.

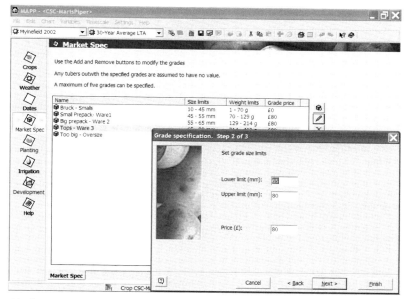

Figure 5. Market specification of a crop. Setting upper and lower size limits and value.

Weather data

Data sets
Each crop stored in MAPP has two sets of weather data associated with it. One is for the year that is of interest, typically the current year and so incomplete, and the other is data for a comparator year that is complete at least until the end of the growing season. Normally, that comparator data comprises a set of long term average weather data (LTA), which is representative of the area where the crop is growing. The accuracy

of the simulations of development and yield that MAPP performs are dependent on the type of weather data entered into the system and users are encouraged to provide as much up-to-date weather data as possible from the farm where the crop is growing. Where current weather data is missing from the current year's records, or has not been acquired yet, the system will switch to using the comparator data for running the simulation. If even only one variable is omitted from the current weather for a particular day then MAPP uses the values for all the variables taken from the second set. This is to avoid combining incompatible values.

As already stated, it is normal to use LTA data as the comparator set so that crop development after 'today' is calculated on the basis of 'likely' seasonal weather. However, this is not necessary. Data from any year or source can be used. For example, a data set from a particularly wet or dry year can be used. Then the user is effectively asking, "What if this year is as wet as ... or as dry as ... from now on?".

Sets of weather data can be created in any of several ways: manually, from a text file, or from a spreadsheet file.

Variables
The variables required for each day are: Day, Month, Dry bulb, Wet bulb, Minimum, and Maximum temperatures (C), Soil temperature at 10cm depth at 0900hrs (C), Precipitation (mm), Solar radiation (MJ/m2), and Potential evaporation (mm).

The weather data can be checked automatically using pre-set thresholds and validation criteria, both optionally set by the user. Failure against one of these tests does not stop the data being used but the anomaly is flagged. Then values that are recognized to be wrong can be edited.

All the variables can be plotted (Figure 6) for visual inspection.

Possible motives for using MAPP

Selecting a seed rate

MAPP uses models of tuber multiplication and size grade distribution for this.

The determining variables include cultivar, seed size, seed price, anticipated yield, and market specification. Decision support is provided through a screen containing both a table and a graph (Figure 7).

Along the bottom of the graph, the abscissa for seed rate runs from a low 1 t/ha to a high 12 t/ha - MAPP is trying to cope with most eventualities. The left-hand ordinate is scaled for total yield and the coloured bands in the graph correspond to the grades specified at Market Spec. They show the yield in each grade varying with seed rate. The table includes a key to the grades. - Click in the graph and a vertical red line appears and the table fills with numbers corresponding to the readings from the graph at the location of the red line. The first column in the box table holds the values given for each grade, the second holds the yield in the grade, read off the graph, and the third column is the product of the first two - value x tonnes. Move the red line with mouse or cursor

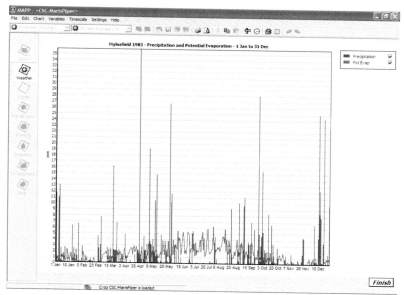

Figure 6. Chart of weather variables.

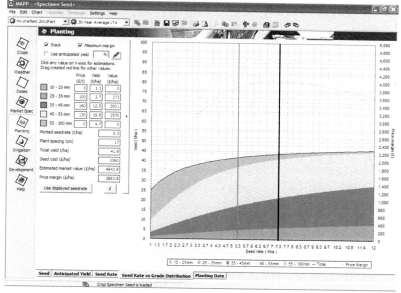

Figure 7. Selecting a seed rate. The black (right-hand) vertical line indicates Maximum Margin, the red (left-hand) vertical line explores the consequences of another seed rate - data in table on the left.

keys and the numbers in the table will change. The margin between the value of the graded crop and the cost of the seed is shown as an orange curve scaled on the right-hand ordinate.

The other boxes show seed rate (where the red line is), plant spacing at that seed rate (considering the row system, specified elsewhere), cost of seed, crop value, and the margin of the value of the crop over the cost of the seed.

Moving the line, the user can explore the likely effect of seed rate on graded yield and on the likely value of the crop at various seed rates.

Optionally, MAPP will display the seed rate corresponding to the highest value of margin (the orange line), but it does not recommend that rate. After all, that point might mean planting a higher seed rate (higher cost) for only a small increase in margin.

In the example shown in Figure 7, the maximum margin, £4043 / ha, occurs at a seed rate of 7.3 t/ha (black line), costing £1460 / ha. The lower seed rate being explored (red line) shows that a seed rate of 5.3 t/ha, costing £1060 / ha, would give a margin of £3884 / ha. Will it be worth spending an extra £400 / ha on seed to recover that money and increase the margin by £159 / ha? The user must decide.

MAPP makes no suggestions. It simply lets the user see how margin of crop value over seed cost is likely to change with seed rate then the user chooses a seed rate to use with this crop in the light of this information and any other factors that he or she thinks are important.

This is the crux of decision support for the user as opposed to having the system make the decisions itself that the user can only accept or reject.

On the same screen the user can test what will happen if the achieved yield is different from that anticipated at planting time. If there is a prospect that the achieved yield will be lower, say, than the value used in the first exercise then the user can readily see the effect on margin and also how the optimum seed rate is affected.

Of course, the user can enter a seed rate determined quite independently of MAPP. MAPP accepts the decision and proceeds to calculate subsequent performance on that basis.

Once the user has selected a seed rate, MAPP provides an estimate of the likely daughter tuber population given the cultivar, size of seed, and plant spacing (Figure 8). It is important to note, however, that MAPP will not use that figure itself. The reason is explained shortly under 'Making sense of test digs' and 'Estimating grade distributions'. Similarly the user can explore the effect of changing the seed size or of growing for a slightly different market (through Market Spec.) - and even the cultivar.

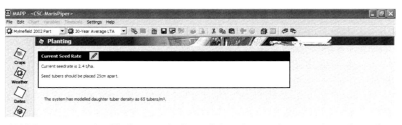

Figure 8. MAPP suggests the likely population of daughter tubers at the selected seed rate.

So! MAPP incorporates the effects of cultivar, seed size, anticipated yield, and market specification (sizes) to show how the yields and TSD in the resulting crop will change with seed rate. And it uses seed price and market specification (prices) to show how the balance between outlay and return changes as seed rate is changed. The user can then make an informed decision.

Examples to illustrate the quantified influence of factors modifying the optimum seed rate

Clearly, the interactive nature of the responses to input data and the rapid presentation of information and options cannot be shown in a printed text. However the following tables (Tables 1 - 3) have been prepared from the output of **MAPP** to illustrate how readily the user can follow the likely outcomes of each choice that is made.
The market specification - the prices to be paid for particular sizes of tuber - is particularly important in determining the return from a crop (Table 1). Here, for a given size and price of seed and anticipated yield, the seed rate that leads to the maximum return over the cost of the seed is shown to change with the price structure for the harvested crop.

Table 1. The influence of market specification on the seed rate giving maximum return.

Valuable size fraction	Value of size fraction			Seed rate giving max. margin (t/ha)	Cost of seed (£/ha)	Max. margin (£)
	30-40 mm	40-50 mm	50-100 mm			
Small	£120	£60	£25	5.6	£980	£1451
Level	£100	£110	£120	4.8	£840	£5485
Big	£25	£60	£120	3.2	£560	£4625

Cvar: Premiere, Size of seed = 650 / 50 kg; Price of seed = £175 / t; Antic. Yield = 55 t/ha; Value of 10 - 30 mm = £0

MAPP can help the grower to accommodate changes in the market that occur during the growing season by changing the dates of haulm destruction and harvest.
The seed rate for the greatest margin over its cost changes non-linearly with seed price (Table 2) and also leads to a non-linear change in the expected number of daughter tubers.

Table 2. The influence of the price of seed on the seed rate giving maximum return and on the associated population of daughter tubers.

Price of seed (£)	Seed rate giving max. margin (t/ha)	Cost of seed (£/ha)	Max. margin (£)	No. of daughter tubers / m²
£80	7.0	£560	£6028	79
£175	4.8	£840	£5485	66
£250	4.0	£1000	£5155	59

Cvar: Premiere, Size of seed = 650 / 50 kg; Antic. Yield = 55 t/ha; Values of size fractions in daughter crop as in 'Level' in Table 1.

The seed rate for maximum return over the cost of the seed is strongly dependent upon the size of the seed (Table 3) which leads to differing expectations of the populations of daughter tubers. Where a seed lot comprises a wide range of sizes, MAPP provides the means to quantify the advantages of split grading the seed and it calculates the best seed rates for each resulting fraction. So, Table 3 shows the contrast between the seed rates for a whole seed lot and its small and large fractions.

Obviously, seed rates differ with the choice of cultivar to be grown. **MAPP** considers the differences in tuber shape and multiplication rate that drive the differences in seed rate between cultivars (not shown).

All these variable inputs and outputs illustrated in Tables 1 - 3, and some others, can be explored freely through the interfaces provided through the icons for 'Market Spec.' and for 'Planting'.

Tracking dates & crop development: MAPP presents the significant dates in the life of the crop in a calendar screen (Figure 9) with scroll boxes to calendars for specifying dates. The key dates identified by the user are those of planting, observed emergence

Table 3. The influence of the size of the seed on the seed rate giving maximum return and on the associated seed spacing, seed cost, and population of daughter tubers.

Size of seed	Spacing (cm)	Seed rate giving max. margin (t/ha)	Cost of seed (£/ha)	Max. margin (£)	No. of daughter tubers / m²
650 / 50 kg (35 - 55 mm)	18	4.8	£840	£5485	66
1205 / 50 kg (35 - 40 mm)	12	4.0	£700	£5750	71
433 / 50 kg (50 - 55 mm)	24	5.3	£927	£5329	63

Cvar: Premiere; Price of seed = £175 / t; Antic. Yield = 55 t/ha; Values of size fractions in daughter crop as in 'Level' in Table 1; Ridge spacing = 90 cm.

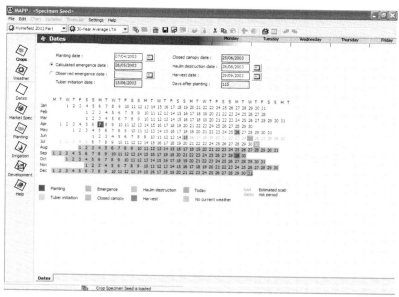

Figure 9. Date screen showing key dates, calculated and observed, the period of greatest risk from common scab, and the switch from current weather to LTA weather.

(optional), haulm destruction, and harvest (optional). MAPP calculates expected dates for emergence, tuber initiation, and canopy closure. In addition, if the crop is to be irrigated for limitation of common scab then the dates for three weeks following the date of tuber initiation are coloured red to indicate when action may be taken to limit the incidence of the disease. During a growing season, real current weather data can only be provided until 'today'. For all subsequent dates, MAPP will use weather data from the second weather file and these dates are shaded grey on the date screen (Figure 9). Under the icon for 'Development', there are tabbed screens for date of emergence, canopy expansion, tuber sampling, yield and graded yield.

Canopy expansion

There are two means to follow canopy expansion. The first, using the 'Canopy expansion' tab, is intended to allow the user to monitor the correspondence between the virtual and real crops. Rather than give values for calculated leaf area index, MAPP shows nine small monochrome photographs (Figure 10) of a canopy expanding from LAI = 0.41 to LAI = 3.45.
As MAPP calculates that the crop has reached one of these stages, the corresponding photograph shifts into colour (Figure 10). Clicking on any one of the nine images produces a larger coloured view for better comparison with reality.
The second means of tracking the canopy development of the virtual crop is available as an option under the tab for 'Total yield' (Figure 1) where a free-standing curve scaled on the right-

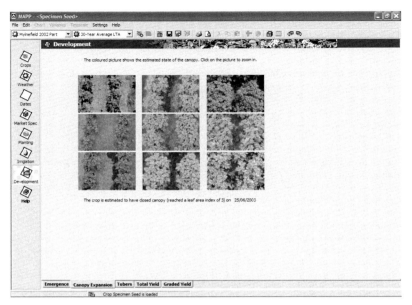

Figure 10. Canopy expansion.

hand ordinate shows simulated senescence as well as growth and is intended as an advanced facility to allow the user to interpret changes in yield towards the end of the season.

Predicting yields

Total yield is simulated using an implementation of Version 3 of the SCRI model of water constrained yield (MacKerron & Lewis, 1995; Rijneveld, 1997) written in Visual C++. Canopy expansion is simulated following Jefferies & Heilbronn (1991), with water supply modulated by root growth and soil water status (MacKerron & Lewis, 1995). Root death and canopy decline is simulated following MacKerron & Lewis (1995) and Rijneveld (1997). Total dry matter fixed is calculated from intercepted incident solar radiation (Jefferies & Heilbronn, 1991); the proportion of total dry matter that is in the tubers (Harvest Index) is determined as a function of thermal time Jefferies & Heilbronn (1991), and the dry matter concentration as a function of thermal time and soil moisture (Jefferies *et al.*, 1989). Weather data from the 'current' set is used to drive the simulation except where there are data missing. For the days where any data is missing from the current data set, data is taken from the comparator weather set (normally the LTA data set). The transition(s) to that second weather set are indicated in the 'Dates' screen by grey shading. Similarly, in the screens showing progress of yield (Figures 1 and 12) the background is shaded grey on those dates.

Either set of weather data can be substituted by another at any stage in the life of the crop and in the running of MAPP. In Figure 1, the smooth progress of the graph of yield after 31st July is because, thereafter, LTA weather is being used to simulate yields. The user can substitute data from another year (wet, dry, hot, or cold) to see the consequences of such weather henceforth. Similarly, a user wishing to examine the consequences of some management operation can substitute several years' weather successively in the current weather set to test the range of outcomes.

Making sense of test digs

It is generally recommended that a grower should make test digs of his or her crop to monitor its progress towards harvest. For users of MAPP this is an absolute requirement if the full capabilities of MAPP are to be used as harvest approaches.

The tuber size distribution (TSD) is strongly dependent upon the population of daughter tubers. Although MAPP gives an estimate of the likely value of that population when the seed rate is chosen (Figure 8) there are many factors that can subsequently modify that figure. These include soil moisture, size of canopy, and intensity of solar radiation, all at around the time of tuber initiation. Further, certain diseases such as Black scurf, caused by *Rhizoctonia solanii*, can modify the final number of tubers that will grow and be harvested. For all these reasons and more, it is better that the grower should check on the real crop rather than relying on a virtual one.

Therefore, MAPP uses the grower's real sample data in order to set the crop tuber population for the graphs of graded yield. The grower has to provide data from at least one test dig, but preferably from several. Obviously, the more accurate and representative the sample data is, the closer to reality the graphs of graded yield will be. The data required for each sample is an identifier for the sample, the length of row sampled, the number of tubers in the sample, and the fresh weight of the sample.

MAPP summarises all the data on number of tubers so as to provide an ever-better estimate of tuber population with successive test digs (Figure 11). In its summary, MAPP presents the grand mean of tuber population together with the maximum and minimum values observed. For the estimate of observed yield, only the data from the latest samples are summarized.

Estimating grade distributions

Once a test dig has been made and one has an estimate of the real population of tubers in the crop, MAPP can be used to examine the development of the TSD as the yield increases. (Figure 12) To view total graded yield, the user must choose a tuber density to supply to the yield simulation. One has the choice of the average, maximum or minimum tuber density calculated from the total of all samples or, to offer complete flexibility, any other arbitrary value can be used. The user can switch between tuber populations to see the significance of the effect and the importance of providing a good representative figure.

On the screen represented at Figure 12, one can see total and graded yields changing with date. As in other screens, the graph is read out into a table (floating, in this case) from a red line that can be moved by cursor or mouse so that the user can easily put numerical values to a position on the graph, i.e. a date. The floating table also shows

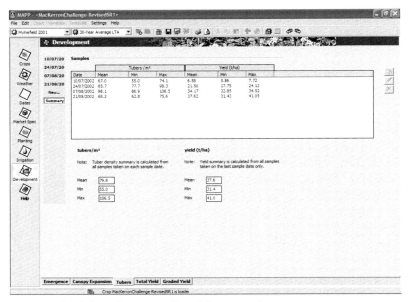

Figure 11. Summarised data from test digs of four occasions.

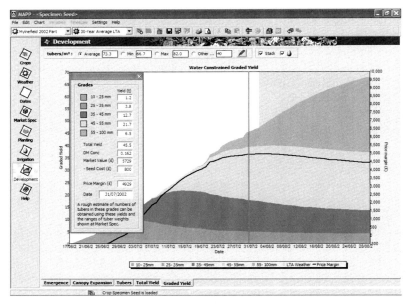

Figure 12. Graded distribution of yield. Note the grey background denoting the shift from 'current' weather to LTA weather data.

some additional data such as dry matter concentration. This can be an important feature in matching a crop to its intended market.

The free-standing black curve shows the margin of the crop value over the cost of the seed and, in the example shown - a seed crop with the smaller tubers having the highest value - it reaches a maximum in the first week of August. Where the market specification gives equal or higher values to larger tubers, that margin increases with increase in yield. The floating table shows all of crop value, seed cost, and margin.

Using this facility within MAPP the user can see whether or not the date of haulm destruction is critical to the grade distribution and the value of the crop and can decide when to kill the crop.

If market conditions should change between planting date and haulm destruction - different prices and different sizes being favoured - then **MAPP** can accommodate this. One simply has to re-specify the Market Spec and all the subsequent valuations are re-calculated. However, the only management option remaining to the user to adjust the grade distribution and the value of the crop is to adjust the date of killing the haulm. Where the revised date would be earlier, **MAPP** offers a check on tuber quality by indicating dry matter concentration.

Scheduling irrigation

The user can indicate that the crop will be irrigated if necessary (default is not to irrigate). Then the user is asked to specify an irrigation strategy (Figure 13): to irrigate to limit common scab or only for growth; what threshold SMD to use for each stage, scab limitation and growth; and what quantity to apply in each simulated case. Additionally the user can request a wider or narrower period for warning to limit common scab. Actual applications of irrigation water are logged - date and quantity - and the user can elect for MAPP to simulate future requirements for irrigation.

Where the LTA weather data is used as the comparator data set then there is always a small amount of rainfall projected for each day and fewer applications of irrigation will be indicated than if the remainder of the growing season is dry (Figure 14). By providing a range of weather data from several years, or by creating artificial data sets, MAPP can

Figure 13. Selecting an irrigation strategy.

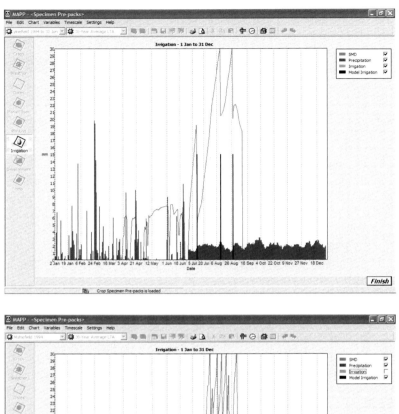

Figure 14. Simulated development of SMD and simulated irrigation based on strategy in Figure 13. (A) with 1994 data to 30th June and LTA thereafter. (B) with actual 1994 data throughout. (A) indicates 3 applications of irrigation may be required. (B) shows that 6 applications would have been called, in the event.

be used to explore the likely demand for irrigation during any particular set of rainfall conditions. So, Figure 14a shows the projected need for irrigation in a particular crop as at 30th June 1994. The simulation uses LTA data from that date, allowing about 2mm of rainfall per day, and projects the need for 3 applications of irrigation - one at 18mm SMD to limit Common Scab and two at 30 mm to sustain growth. In Figure 14b the simulation has been given the data for all of 1994 and so it shows how demand for irrigation would have worked out in the event, with six applications of irrigation being required.

Where future irrigations are simulated their water is added to the supply available to the crop in the future and its effect on canopy growth, yield etc. is included in the simulation (Figure 15a, b). As the dates of these future irrigations pass, then they are replaced by actual weather and actual irrigation practice. Figure 15a shows the development of leaf area index and total yield, driven by the actual weather of 1994. The leaf canopy expands but the canopy is closed (LAI > 3) for only a short period, and yield is limited to 42 t / ha. Where irrigation is applied, as simulated at Figure 14b, then the leaf canopy expands to LAI = 4.7 and the yield is simulated to reach 58 t / ha (Figure 15b).

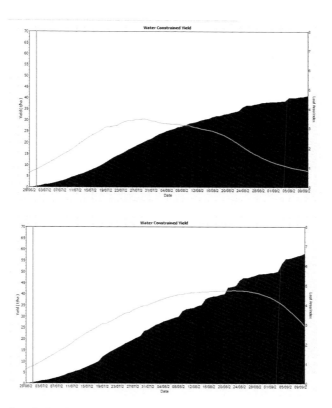

Figure 15. Simulated development of yield and LAI using weather data of 1994, under unirrigated (A) and irrigated (B) conditions. (B) shows the simulation if the scheduling indicated at Figure 14b were to have been applied.

What if ....?

Most computer-based systems allow input of multiple variables - even expect them - but generally they give single answers. Most such systems give no indication of the confidence to be placed in the result or of the sensitivity of the answer to the changing values of an input and, therefore, give no indication of what will be the result of a differing input. These are two related but separate contributory factors in making a decision that should be contained in any DSS. MAPP does not show standard errors or confidence limits but it does show the sensitivity of certain of its answers to changes in the inputs. Thus, Figure 7 shows the calculated effects of changes in seed rate on grade distribution and on the value of the crop as continuous functions. The user can make an informed decision. Similarly, Figure 12 shows the calculated effects of changes in date of haulm destruction on grade distribution and on the value of the crop as continuous functions.

Should similar displays be provided for all sets of input variables? Probably that would not be useful. As things are arranged in MAPP, the effects of input variables that can be changed step-wise, e.g. cultivar or size of seed, can be examined by the user in a step-wise fashion. It is the variables that can be changed continuously e.g. seed rate and date of haulm destruction for which there is the continuous display. Among the tools that MAPP offers, apart from those just discussed, is the ability to change critical dates and examine the consequences. It also allows the user to examine the effect of changing irrigation strategy, e.g. if water is limited, what are the effects of delaying the start or of changing the phasing of the applications.

Interaction with users

Beta version and beta workshop

In the course of development of MAPP, both the concept and the details were presented repeatedly to growers and agronomists and their comments and suggestions were considered and some were adopted to modify the development. After two years of development, in May 1999, there was a 'working' version available (the 'Beta Version') and this was presented to a panel of growers and agronomists. Sixteen people were given a presentation about MAPP then given a 'hands-on' session with the package before being given a copy to use on their own computers, at their own convenience. We called this, 'Beta Testing'. Again, comments and suggestions were sought and where possible the ideas were incorporated into MAPP, particularly those that had a bearing on the appearance and comprehensibility of the package.

These tests were coupled with exhaustive checking of all the links and processes in the package. Among the faults that were found and corrected were: New crops not located intuitively, problems assigning weather data, insufficient categories of soil, sloppy arithmetic in one or two places e.g. in conversion between acres and hectares. These and many other problems were ironed out before MAPP Version 1 was launched in

February 2001. The involvement of people who are representative of the intended users was an essential part of the development of the package.

Problems experienced since launch

Of necessity, MAPP generally has to be installed by the users themselves and there is a wide diversity of machines and operating systems in use. There were initially a few instances of difficulties with installation. The only one that could be attributable directly to the way that MAPP was written was the one that requires the Regional Settings in Windows to be set at 'English (UK)'. This probably reflects on the way that dates are handled but also, possibly, the symbol used as the decimal point.

Owing to the careful and exhaustive testing, few if any faults were found in the operation of MAPP itself. However, there were instances when its 'intuitive' operation was evidently not sufficiently intuitive and some users needed to be helped. Still others were not aware of its full functionality until shown in a demonstration.

An 'Illustrated Guide' was prepared using PowerPoint, rather than a printed manual, and was included on the installation disk. This enabled a free use of screen shots and targeted explanations. However, experience suggests that even this feature is not used by those people who need help.

These characteristics of general users point to the need for workshops or one-to-one tutorials to acquaint users with a new package.

Weather data

MAPP needs weather data, ideally from weather records taken from close to the crop in question, to calculate crop development and yield through the year up to the present, whenever that is. Then, to carry its calculations forward towards haulm destruction, MAPP must have a file of weather data that extends that far. Normally MAPP uses a file of long-term averages (LTA) adopting the idea that an LTA represents the weather that the crop is as likely as not to experience in the future. MAPP has been provided with several files of weather data, selected to represent several types of weather - wet, dry, cool spring, warm spring, 'typical' etc., and also some LTA data to help users get started. Users found three separate problems. One was to obtain LTA weather data for their region. We provided LTA data for Scotland and five regions of England. The second problem was to obtain current weather data. The ideal is to use or have access to an automatic weather station but for those who did not have that resource, obtaining current weather data was a problem. The third problem hinged on the need regularly to enter the current weather data. This was not always straightforward for those who were not entirely familiar with handling data on computers.

Results from a users' workshop, July 2002

Attendees volunteered that they had used MAPP for: Irrigation scheduling, Projecting yield, Grade distribution, Market specification, Deciding on date of burn-down. They admitted to having problems with: Adding a new cultivar, Number of tubers, Depth of planting. Other problems included the wish to transfer weather data automatically from weather station to MAPP. MAPP is very self-explanatory but it is very easy for those who are steeped in the use of computers to overlook or to misjudge the difficulties seen by those who use computers less often.

Conclusions

Experience has shown that the most successful use of mathematical models in making tactical decisions is found where field observations (made by the farmer) are combined with computerised systems. MAPP allows the user to enter certain data from the real crop to constrain the virtual one. Indeed, before calculating graded yields, it requires this.

DSS should enable estimates, at least, of the differing outcomes consequent upon the weather differing from expectation. MAPP allows the use of a range of weather data following from 'now', whenever that is, so that the user can be aware of the range in outcomes.

DSS should provide estimates of the consequences of differing actions - at least of action versus inaction or of the comparison between less being given and more. MAPP reveals the consequences of changing continuous variables that are within the grower's control, such as planting dates, seed rates, and date of haulm destruction, as well as discontinuous variables such as choice of cultivar or irrigation strategy.

If a DSS is to be successful it must be accepted by the intended users and so it must address the needs and opinions of those users. That can be achieved only by consulting panels of users, reiteratively, as the system is being designed.

The design of a DSS should pay particular attention to requirements for data, possible sources of that data, and the cost of its acquisition. Weather data is potentially the most troublesome to acquire.

Information required should include limits to the amount of irrigation that is available, prices of inputs, and the value of the output from the crop.

Finally, 'Decision Support' should not be confused with the decision itself. Decision support means giving options, not making the decisions. A key feature of the way in which MAPP operates is that it shows, not only what the system calculates is 'best' but also, the consequences of doing something different. The user can then make an informed decision.

References

Jefferies, R.A., T.D. Heilbronn & D.K.L. MacKerron, 1989. Estimating tuber dry matter concentration from accumulated thermal time and soil moisture. Potato Research, 32, 411-417.

Jefferies, R.A. & T.D. Heilbronn, 1991. Water-stress as a constraint on growth in the potato crop. I. Model development. Agricultural and Forest Meteorology, 53, 185-196.

Jefferies, R.A., T.D. Heilbronn & D.K.L. MacKerron, 1991. Water-stress as a constraint on growth in the potato crop. II. Validation of the model. Agricultural and Forest Meteorology, 53, 197-205.

MacKerron, D.K.L. & P.D. Waister, 1985. A simple model of potato growth and yield. Part I. Model development and sensitivity analysis. Agricultural and Forest Meteorology, 34, 241-252.

MacKerron, D.K.L. 1987. A weather-driven model of the potential yield in potato and its comparison with achieved yields. Acta Horticulturae, 214, 85-94.

MacKerron, D.K.L., D.J. Greenwood, B. Marshall, R. Rabbinge & B. Schöber, 1990. Forecasting systems for the potato crop. In: Proceedings of the 11th Triennial Conference of the European Association for Potato Research (ed D.K.L. MacKerron), pp. 85-106. EAPR, Edinburgh.

MacKerron, D.K.L. & G.J. Lewis, 1995. Modelling to optimize the use of both water and nitrogen by the potato crop. In: Ecology and modelling of potato crops under conditions limiting growth (eds A.J. Haverkort & D.K.L. MacKerron), pp. 129-146. Kluwer.

Marshall, B., J.W. Crawford & J. McNicol, 1995. Handling qualitative and uncertain information. In: Potato ecology and modelling of crops under conditions limiting growth (eds A.J. Haverkort & D.K.L. MacKerron), pp. 323-340. Kluwer Academic Publishers, Netherlands.

Marshall, B., 2000. MAFF Final Report for Project Code HP0210T. A predictive model of potato size distribution and procedures to optimise its operation, 1997 - 2000. Ref Type: Report

Marshall, B. 1986. Tuber size. Aspects of Applied Biology, 13, 393-396.

Rijneveld, W., 1997. POTATO_3. Finishing, Testing, Analysing and Validating a Decision-supporting Simulation Model. 42 pp. MacKerron, D. K. L., Invergowrie, Dundee, SCRI. Ref Type: Report

Sands, P.J. & P.A. Regel, 1983. A model of the development and bulking of potatoes, (*Solanum tuberosum* L.). V. A simple model for predicting graded yields. Field Crops Research, 6, 25-40.

Smart, J.A.C. 1992. Model Abstraction: An A.I. Approach to Flexible Crop Modelling. PhD University of Dundee.

Name of the DSS

Integrated Management Strategies for Potato Cyst Nematodes

Who owns the DSS, principal author's address,...

Martin Elliott, Scottish Crop Research Institute, Invergowrie, DD2 5DA Dundee, United Kingdom
mellio@scri.sari.ac.uk

What is it for, what questions does it address?

- Management of PCN *Globodera pallida* in ware potato fields
- Shows changes in population densities over 5 rotations after potato crops
- Shows total yield expected at harvest from PCN infested land (5 rotations)
- What is a sustainable cropping strategy?
- What effect does cultivar resistance and tolerance have on population density?
- What effect do chemical treatments have on PCN populations?
- What are the combined effects of resistance, tolerance, treatment and rotation on PCN population densities and yields?

What input is required?

- Models can be customised for specific fields - pre-planting and related post harvest population estimates from across the field along with yield estimates from the same points.
- For known fields, just pre planting population estimate necessary.
- A matrix of pre planting population estimates can also be used

Who is it for, who are the intended users?

- Agronomists
- Farmers
- Teachers

What are the advantages over conventional methods (whatever those may be)

The main controllable components required for PCN management can all be manipulated. The consequences of varying any of the inputs over five potato crop rotations can be see immediately.

How frequently should the DSS be used or its values be updated?

Can be used at any time

What are the major limitations?

- Extensive sampling required to get sufficiently accurate data to parameterise the models for specific fields.
- Knowledge of potato tolerance and resistance to *G. pallida* and how this can vary.

What scientific / technical enhancements would be desirable?

9. Projecting PCN population changes and potato yields in infested soils

Martin J. Elliott, David L. Trudgill, James W. McNicol and Mark S. Phillips

Models describing the relation between population sizes of the potato cyst nematode *Globodera pallida* at planting and harvest and yields from infested soils have been incorporated into a computer program to assist growers manage the PCN populations in their fields. In the absence of commercial potato varieties able to confer full resistance to *G. pallida*, it has been found in practice, through observed population density increases, that there is no single method of control. However, control is achievable by combining several measures. The program assists with management strategies for any field, but also includes in its database the trial sites used to generate the underlying models. The user can investigate the effect of changing variety tolerance and resistance, the length of rotations and the efficacy of any chemical treatment. The program can use farm-produced data to customise the models and extend the database. This has proved to be a valuable tool for UK potato growers to help them understand the interaction of their cropping strategies and the possible consequence on both future PCN populations and yield expectations.

Introduction

Potatoes were first introduced to the UK from their native South America as early as 1570 (Evans *et al.* 1975) and were probably followed some time later by the very damaging plant parasitic potato cyst nematodes (PCN) *Globodera rostochiensis* (Woll) and *G. pallida* (Stone). The most recent survey of potato growing land in England and Wales showed that PCN was present in 64% of sites sampled and that the populations were distributed in the proportions 67% *G. pallida*, 8% *G. rostochiensis* and 25% both species (Minnis, 2002). This verified the shift observed in the ratio of nematode populations that, before the 1960s, was predominantly *G. rostochiensis* but is now *G. pallida*. This change occurred after the introduction of potato cultivars with complete resistance to *G. rostochiensis* but susceptibility to *G. pallida*. Growing these cultivars resulted in the selection of *G. pallida,* previously present at often undetectable levels. In populations of mixed species, where doubly-susceptible potatoes are grown, *G. pallida* is out-competed by *G. rostochiensis* as the former generally hatch later. The new cultivars allowed the low population densities of *G. pallida* to increase at a time when growers believed that their nematode problems were over. Now, modelling has shown that under certain conditions, insignificant populations can become damaging over 6 rotations (Figure 1).

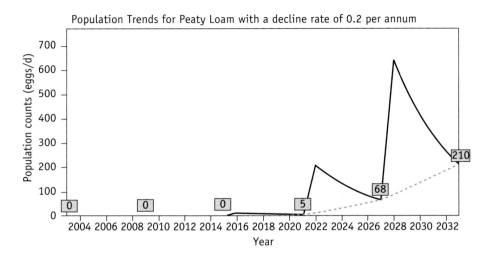

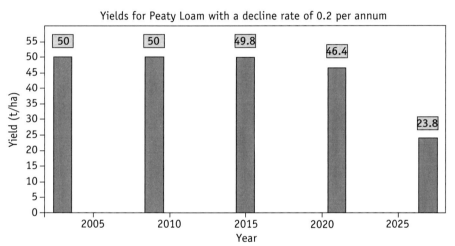

Figure 1. Very low population densities of PCN Globodera pallida (0.001 eggs/g soil) rise rapidly when exposed to susceptible potato varieties on commercial rotations (six years), with no nematicide treatment, to become damaging epidemics in a relatively short time..

As the area devoted to growing potatoes in the UK has decreased and become more specialised (Minnis, 2002), high yielding, tolerant varieties have been developed which incorporate resistance to *G. rostochiensis* preventing the re-emergence of the pest but, again, assisting in the increase of *G. pallida* populations. Currently the only commercial cultivars available for controlling *G. pallida* have only partial resistance and, in the UK, have limited market presence and represent only a small proportion of the area grown. However these partially resistant cultivars can be an important component of an integrated management strategy for controlling PCN. The other control strategies are long rotational gaps between potato crops and the use of nematicides.

To explore the consequences of specific PCN management strategies a computer program has been produced to simulate potential loss of yield in the presence of PCN, and to illustrate the effects of possible control measures on the population dynamics of the nematodes over a number of cropping cycles. The program is based on two models, one of nematode population dynamics and the other of proportional yield loss.

Development of the models

The mathematical models used in this program were constructed from data on pre-planting and post-harvest populations of G. pallida and yields obtained from field trials carried out at four sites. Two were in England and two in Scotland, on differing soil types with cultivars of differing resistance and tolerance characteristics to G. pallida. The plots were pre-treated by planting cultivars, in the previous year, with varying degrees of resistance and tolerance to G. pallida to create a wide range of pre-planting densities for the main trial.

Three models are used. The first relates yield to initial nematode population (P_i) and the second relates post harvest nematode population level (P_f) with initial population level. The third is used to estimate the annual decline in the number of nematodes surviving in years when non-host crop is grown. This is assumed to be a constant.

Yield loss

Yield in relation to initial nematode population density is determined using an inverse linear model (Elston et al., 1991):

$$E[Y] = Y_{max}/[1+(P_i/c)] \qquad (1)$$

Where,
E[Y] represents the expected yield,
Y_{max} is the expected yield in the absence of nematodes,
c is a measure of tolerance of the plant to damage by the nematodes invading the roots. This parameter is partitioned into a site and genotype component.

Population dynamics

A logistic population limitation model (Seinhorst & den Ouden, 1971) has been used to model population dynamics. This takes the basic form.

$$P_f = M(1-exp[aPi/M]) \qquad (2)$$

Where
P_f is the final population after harvest,
M is a theoretical maximum population density in the absence of damage to the plant,
a is the maximum multiplication rate, measured at low population densities.

Equation (2) is modified in a number of ways. In particular, account has been taken of damage to the root system following nematode invasion both in terms of the amount of root available for nematode invasion and reproduction and also in relation to the degree of hatch of nematode eggs. Not all nematode eggs hatch and a proportion of the initial population remain in the soil and may hatch when subsequent non-host crops are grown. Consequently an expression that accounts for this has been added to Eqn. (2). The model also includes a parameter to account for nematicide efficacy.

The Decision Support System (DSS)

Data requirements

The program includes a database as well as the models. The database is organised around farms and their associated fields and allows the recording of field cropping information together with the parameters used by the models.

Strategy assessment estimates
To assess a strategy, the user needs first to identify the farm and field then input an estimate of P_i or a grid of P_i estimates. Cropping details are required for the following: cultivar tolerance - for which there are 4 categories see Table 1; cultivar resistance on a scale from 1 (fully susceptible) to 9 (fully resistant), length of rotation between potato crops, the expected nematode-free yield, and any nematicide treatment. Initially this information is applied to all subsequent rotations. Trend charts for the population and yield for the next five potato crops are produced automatically (Figure 1). Optionally, crop details can be changed for each rotation.
Graphs of the relation between yield and P_i, and between P_f and P_i, are also produced for the cultivar or cultivars used in the assessment.

Table 1. Tolerance categories and cultivar examples used in the program.

Tolerance category	Example cultivar
Very intolerant	Maris Peer
Intolerant	Maris Piper
Tolerant	Kerr's Pink
Very tolerant	Cara

Customising the model

The program has default parameters for three different soil types based on the initial studies mentioned above. However a user can estimate parameters for a field if a set of data is available that consists of pre- and post-harvest nematode population densities and yields. The range of P_i values should be as wide as possible and the soil composition across the sampled area needs to be relatively uniform. The parameters in the yield loss model are estimated first, followed by those for the population model. These estimated soil parameter values are stored in the database ready for future assessments of strategy.

Data quality

Precision and accuracy of estimates of nematode population levels are important. Currently the sampling strategies for ware potato fields are changing as the incidence of PCN and damage they cause the crop increases. Typically, the majority of laboratories in England would process 100g of dried soil, sub-sampled from a larger sample taken from blocks of 4 ha. This strategy can miss patches of PCN up to 0.1 ha and takes no account of the distribution of PCN population densities within the field. ADAS in England and SAC in Scotland process larger samples and there is a move towards examining smaller areas while maintaining, or increasing, the volume of soil processed. Although this provides an improvement in precision, estimates of infection levels will still be worryingly imprecise.

User interface

The DSS will operate on any modern PC. The program (Figure 2) begins by offering 4 main options, namely:
- Control strategy analysis: Allows the user to investigate the influence of P_i on yield and P_f over successive rotations using different control strategies
- Spatial analysis: Given a grid of P_i values, investigate the spatial influence on yield and on P_f for a single crop using different control strategies or plot a complete grid of P_i, yield and P_f.
- Customise the models: Using P_i, P_f and yield data, customise the models with respect to site tolerance and nematode multiplication.
- Modify database: Allows the user to add new farms or modify existing information.

Within each of the above options there are additional menus with more options but these only become active after data has been supplied. The user is unable to make changes to any of the default settings until data entry is complete.
The additional options include:
- Varying the estimate of initial population density given at the beginning of an assessment.
- Varying the annual decline rate.
- Varying the nematicide efficacy of the treatment applied to any of the crops.
- Producing a report of the assessment, recording the charts and the cropping details used.

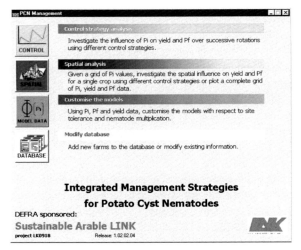

Figure 2. DSS main screen showing the four main options.

- Copying the charts produced to report files.
- Scaling the population and yield trend charts.

The options for the cropping details are presented in drop-down-lists (Figure 3).

Example

Consider a data set of P_i, yield and P_f from an infested field on a farm previously registered to the program, in which a susceptible potato crop was grown. The field will have to be registered with a unique reference so that the details can be recalled. The data can be either pasted from another application or accessed directly from a file. When the data is entered into the program, the yield is plotted against the P_i with an adjustable curve produced by the model (Figure 4). The tolerance of the cultivar is first identified from a drop-down list. The curve is then manipulated with slider controls to vary the nematode-free yield and the site tolerance parameters. The model is fitted interactively to the data. The quality of the fit is measured by the sum of squared

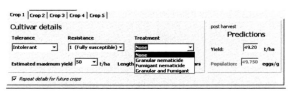

Figure 3. Crop details are obtained from drop-down-lists. In the first instance, only the details for Crop 1 will be visible, the others become accessible when Crop 1 details are complete.

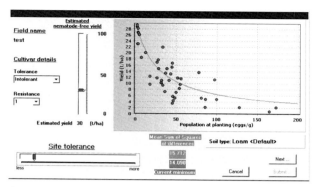

Figure 4. Window for customising the yield model for specific fields using P_i, P_f and Yield data with the fitted yield loss model plotted through the data. The sum of squared differences between the fitted and actual data values is used to assess the quality of model fit. Current and best-to-date values are shown.

differences between the actual and fitted values. The minimum that has been achieved during the fitting process is displayed below the current value (Figure 4). Once the user is satisfied that the "best" curve has been found for the yield data, the population data is plotted, P_f against P_i. The same process of finding the best fit of the population dynamics model is achieved by adjusting the nematode multiplication rate slider (Figure 5). When the fit has been optimised, the user can accept that the fitting is complete and commit the parameter values to the database.

Any data point can be highlighted in these graphs to reveal its actual values, and if required, individual points can be identified for exclusion from the model fitting process. The exclusion of a point is indicated by a change in colour. A point so removed from the yield data is also identified when the population data is presented. These points can be reintroduced into the analysis by clicking on them and agreeing to their re-inclusion.

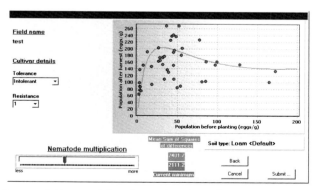

Figure 5. Fitting the population model to data using the soil tolerance parameter obtained from the proportional yield loss model.

Standard errors for the parameter estimates are not provided. This is because of the difficulty of estimating the sizes of PCN populations. The extent of this variability depends on the time since the initial introduction, the number of subsequent potato crops, the time between the crops and the amount of cultivation that has taken place. With the parameter values committed to the database, the field can then be recalled for strategy assessments.

Strategy assessment based on spatial data can also be obtained for a field if a grid or matrix of P_i estimates is available. In this case, the assessment is for a single crop and all P_i values are plotted on a single chart using triangulation to build a surface of P_i levels. When the crop details are entered, the program will plot a similar chart for both the yield and the P_f, based on information already given.

Future developments

There are several possibilities for improving the accuracy and scope of the program. Improvements could be made in accuracy by further studies of the annual decline rate. This is the natural decrease in the population due to the death or removal of cysts, eggs and juveniles. PCN have an extremely narrow host range and can multiply only on potato, tomato and other solanaceous plants. In the absence of these hosts, populations can decline due to natural, spontaneous hatching, predation from insects and fungi, and the physical damage in the soil by machinery or other mechanisms. The rate of decline appears to vary between sites (Turner, 1996) and the DSS predictions are sensitive to changes in the decline rate. Hence the importance of accurate estimations being obtained. This can be achieved by sampling from the same area of the field over many years. However, there is an economic cost involved that will restrict many growers from undertaking such an exercise as they are already facing financial hardships. It may be possible to collect enough information by sampling from the same locations from two consecutive crops to get an estimate of this parameter without incurring high costs.

Laboratory experiments have shown that repeated growing of partially resistant potato genotypes can select *G. pallida* populations for increased virulence (Turner, 1990, Turner & Fleming, 2002). The DSS currently has no facility for modelling selection for virulence in the scenario where partially resistant cultivars are used intensively in a rotation. Differences in virulence of *G. pallida* populations cannot be directly manipulated other than by altering the resistance of the cultivar. Further studies of field populations over a number of cropping cycles will be required to obtain adequate data to measure these effects.

User base

The current user base is centred on agronomists who are undertaking trials for the British Potato Council (BPC) or are contributing to a Sustainable Agriculture LINK project entitled "Integrated management strategies for potato cyst nematodes". In addition, and within this project, there are research institutes (Rothamsted Research and Harper Adams University College) that are able to use the program as an educational aid. The program has been extensively demonstrated to grower groups and to agronomists. In view of the complexity of the PCN problems it is advisable that any user will have received training from an experienced field nematologist.

Benefits of this program

The benefit of this program is that it handles the main control methods for PCN and it is easy to manipulate these and immediately see the effect of the control in combination or in isolation. The program illustrates graphically the principles underlying the changes in nematode population levels and the consequent effects on yield. Growers can be alerted to the potential problems of not managing PCN population growth. If there is already a problem the grower can obtain an indication of the level of control of nematode populations resulting from the different control strategies. Local and possibly low quality data can be used to parameterise the models to produce strategy assessments specific to particular fields or environment types.

Another obvious advantage is that with the program, growers can be educated to understand their own PCN problems and take the steps necessary to begin to control it or to prevent it becoming a major problem. There are many examples of the dire consequences of ignoring PCN. Populations levels will rise and quickly render land unusable for potato production.

An additional feature of the DSS, as with any modelling package, is that the consequences of assumptions not holding can be investigated. For example, it has been assumed that granular nematicides are 80% effective at controlling PCN but in reality, their effect is extremely variable as they have a very narrow operating range and are extremely susceptible to environmental conditions. With the DSS the user can vary the effectiveness of the nematicide and so observe the consequences for his control strategy when the efficacy is not 80%. Without the program, these consequences can only be assessed through retrospective sampling.

References

Elston, D.A., M.S. Phillips & D.L.Trudgill, 1991. The relationship between initial population density of potato cyst nematode *Globodera pallida* and the yield of partially resistant potatoes. Revue de Nematologie 14, 213-219.

Evans, K., J. Franco & M.M. De Scurrah, 1975. Distribution of species of potato cyst-nematodes in South America. Nematologica 21:365-369.

Minnis, S.T., P.P.J. Haydock, S.K. Ibrahim, I.G. Grove, K. Evans & M.D. Russell, 2002. Potato cyst nematodes in England and Wales - occurrence and distribution. Annals of Applied Biology 140, 187-195.

Seinhorst, J.W. & H. den Ouden, 1971. The relation between density *of Heterodera rostochiensis* and growth and yield of two potato varieties. Nematologica 17, 347-369.

Turner, S.J., 1996. Population decline of potato cyst *nematodes (Globodera rostochiensis, G. pallida)* in field soils in Northern Ireland. Annals of Applied Biology 129, No. 2, pp315-322.

Turner, S.J., 1990. The identification and fitness of virulent potato cyst-nematode populations (*Globodera pallida*) selected on resistant Solanum vernei hybrids for up to eleven generations. Annals of Applied Biology. 117:385-397.

Turner S.J. & C.C. Fleming, 2002. Multiple selection of potato cyst nematode *Globodera pallida* virulence on a range of potato species. I Serial selection of Solanum hybrids European Journal of Plant Pathology. 108: 461-467.

Name of the DSS

NemaMod

Who owns the DSS, Principal author's, address,...

Thomas Been and Corrie Schomaker, Plant Research International, Wageningen University and Research Centre, P.O. Box 16, 6700 AA Wageningen, The Netherlands. Thomas.been@wur.nl

What is it for, what questions does it address?

- What is my risk of a defined loss of yield?
- Should I use a non-fumigant nematicide?
- Should I fumigate the soil?
- What are the cost/benefit ratios of these nematicides?
- Is my choice of cultivars optimal when I look forward several rotations?
- Which cultivars are optimal (field specific) in my situation?
- How can I manage my nematode population so that the risk of detection of Q-organisms by any statutory monitoring system in fields or lots is nil?
- What soil sampling method is the most profitable when managing my nematode population?
- When is the optimal frequency of soil sampling in my situation?

What input is required?

- Sampling data
- Cultivar-specific information, e.g. about resistance to and tolerance of nematodes and other pathogens, and quality
- Profit per unit product
- Costs of control measures
- Geo-data.

Who is it for, who are the intended users?

Education (universities, agricultural schools), Extension services, Farmers and the Agro industry

What are the advantages over conventional methods (whatever those may be)

NemaMod is based on principles of quantitative risk management.

The conventional methods use rough estimations of mean values of parameters, initial population densities and partial resistance and therefore leave too much room for speculation and individual perceptions of risk. The conventional methods offer no cost/benefit analyses.

How frequently should the DSS be used or its values be updated?

NemaMod should be used once a year before planting a new crop of potatoes as *NemaMod* is a strategic DSS. During crop growth no control measures for nematodes are possible. Values can be updated when new research results become available.

What are the major limitations?

Major limitations are gaps in our knowledge of nematode species such as *Meloidogyne* and *Pratylenchus penetrans*

What scientific/technical enhancements would be desirable?

It would be desirable to make *NemaMod* an integral part of a flexible crop protection system connected to precision farming.

10. A geo-referenced Decision Support System for nematodes in potatoes

T.H. Been and C.H. Schomaker

An ActiveX-component called *NemaMod* has been developed containing all available, relevant quantitative knowledge about plant parasitic root nematodes (cyst nematodes, root knot nematodes and *Pratylenchus* spp.) in some major crops, but with emphasis on potatoes. The component is the core of a decision support system (DSS) for nematodes. Because of its nature, an ActiveX-component enables the development of different interfaces by different software developers. The DSS for nematodes will be connected to a Farm Management System (FMS), which contains all farm-related information necessary for agricultural practice, administrative duties required by the Government and certification. Moreover, by way of the FMS the DSS has access to Geo-information for farmers and results provided by statutory soil sampling agencies. Geo-information enables sampling results to be visualised, giving an overview of the on-farm situation on nematodes. The farmer can use the DSS to calculate risks, compare cropping scenarios, ask 'what if' questions on the basis of sampling data and he gets site-specific answers to his questions displayed in the visualization module. *NemaMod* enables risk management for crop losses over a number of cropping frequencies with various levels of stochasticity. The DSS can be applied for both infestation foci and full-field infestations. The primary task of the DSS is to reduce the populations of nematodes to low, economical levels that are not harmful and keep them at these levels. It will facilitate certification harmonised with European (Eurep-GAP) and global standards (GFSI) for food safety and sustainability (promoted by UNEP). *NemaMod* can also act as an educational tool.

Introduction

Due to the EGFL (European General Food Safety Law), which becomes effective in 2005, food safety is becoming an increasingly important issue in the agro-food production chain. The EFGL demands end-to-end tracking and tracing and makes food producers legally accountable for the safety of food products. Consequently, several quality assurance systems have been developed for various links in the food supply-chain, evolving from basic food safety schemes such as HACCP, towards Good Practices in Agriculture (EUREPGAP, BRC), Manufacturing and Distribution. The development of certification schemes for food safety and sustainability is still in its early stages. The ultimate goal: quality assurance systems for the whole food supply-chain is only achieved when all product entities have a unique, shared identity and all participants in the supply-chain have access to a database containing the full history of the product. Therefore, certification of production chains will be a major issue in the next decades and will have global coverage. The first steps towards global certification have been initiated by the CIES (The Food Business Forum). Eventually, wherever in the world a product is certified, it will be accepted everywhere. Next to food safety, quality and sustainability are logical extensions for inclusion in certification schemes. For the primary production sector this implies the registration by farmers of all activities related to crop management. Pesticide applications present food safety issues and, therefore, Integrated Pest Management (IPM) is a major issue in certification schemes. So, decision support systems for pest, crop and farm management will become more and more important.

Because of the necessity of a shared database for the food supply-chain and the need to lessen the administrative burden on farmers, currently estimated at 4 hours per week, the old paper-based filing systems are inadequate. To improve operations, Farm Management Systems (FMS) have been developed with the approach of 'input once, use frequently'. Among the positive results of this development are the rapid growth in the number of farmers having a Personal Computer and using an FMS, and the possibility of using this digital information in DSS to improve farm management.

In the late 1990s, the Dutch government decided to develop the so-called 'Basisregistratie Percelen', a Geo-referenced database containing the polygon of every farmer's field in The Netherlands (750,000) and each crop area within these fields. This database enables the visualization of crops and fields in a given year. The government decided to make the information in this database available to farmers for farm management purposes. By employing this basic information, it becomes feasible to automate the production of charts presenting site-specific information, e.g. the monitoring results of pests, diseases and nutrients.

Plant parasitic nematodes are still an important cause of crop losses in the temperate zone. This especially true of potato cyst and beet cyst nematodes, root-knot nematodes, root-lesion nematodes and *Trichodoris* species. In the past, attempts at their control was relatively straightforward using crop rotations and nematicides success indicated by mean multiplication rates on hosts and non-hosts. However, nowadays this approach to nematode control is neither economically nor socially acceptable. Moreover, it is not in agreement with current scientific views on floating boundaries between susceptibility and resistance. A modern alternative is a quantitative risk management system for nematodes. Because of the complex nature of such a system - it includes several non-linear stochastic models and large databases - it must be embedded in a software environment. This environment must be differentiated for user-groups: e.g. farmers, extension services, policymakers and students. The system must be organised so that important, everyday questions, for instance on the choice of monitoring systems, crop rotation, cultivars and planning of seed potatoes, cost benefit ratios of control measures, how to keep fields and potato lots "free" from harmful nematodes, etc. can be answered directly. Agro-chemicals are only used if preventive non-chemical control measures are inadequate.

In co-operation with government, trade and industry it was decided to disseminate the large amount of quantitative knowledge on several important nematode species that cause problems in Dutch agriculture in a nematological module **NemaMod**. For a start **NemaMod** focuses on rotations with potatoes. To enhance the acceptability and use of **NemaMod** it will minimise data input by farmers. Ideally, farmers only have to ask their questions of the DSS to get a straight answer. Therefore, the availability of data, whether it be the Geo-information or the monitoring results, should be straightforward. To assist the rapid development of FMS, which can function as the data-providers to DSS, it was decided to design **NemaMod** as an auxiliary to these FMS.

It would not be wise to develop DSS separately for the various plant diseases. Each DSS will need some basic functions like visualization of monitoring data using Geo-information and databases with information on legislation, crop characteristics, etc. Therefore, a modular, integrated system for crop protection is envisaged. This system will be used to manage all major pests and diseases and contain separate modules for nematodes, viruses, fungi and bacteria.

The Decision Support System

The complete decision support system comprises the following building blocks:
- An ActiveX-component **NemaMod(r)** incorporating all nematological knowledge, 'the engine'.
- An ActiveX-component **SampView(r)** incorporating logic for sampling and geometry, and algorithms to divide fields into agronomic areas (sampling units, cropping areas, etc.).
- A visualization component.
- Databases increasingly provided by FMS.

Farm Management Systems are the core of the DSS (Figure 1). Input to an FMS will come from different data-providers such as soil sampling agencies, Geo-information, legislation, crop protection information, economic information and the farmer himself. Some of the FMS in operation now, already have some of the required input such as the approval of pesticides for different crops. The effort in the current development is targeted at getting the sampling results and the Geo-information from the data providers to the FMS in a pre-defined format to enable visualization. Building and connecting the **NemaMod** component to a FMS and building the **SampView** component to enable presentation of the sampling results in a general visualization component will be the main effort.

Outline

The main part of **NemaMod** is the ActiveX-component, a custom-control usable in different programming environments, providing the functionality of the DSS. The implementation of the DSS as an ActiveX component has some major advantages. In the

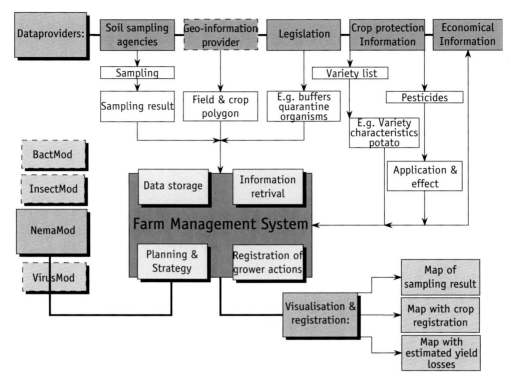

Figure 1. Possible layout of an integrated system for crop protection. Data- providers are linked to a Farm Management System (FMS) and can relay farm-specific data if requested. The FMS is linked to the several modules that constitute the different DSS. At any time, monitoring results, cropping areas and advice can be visualized.

first place, the development of the scientific engine is independent of any user interface. Therefore, different software developers can use the **NemaMod** component to add decision support for nematodes to their own Farm Management System. They can develop their own customised user interface and choose the degree of complexity (e.g. functional character) they want to convey to the user. As communication with the user is handled by the interface the **NemaMod** component causes no language restrictions and can be used by developers all over the world. Further, it both enables the re-use of programming code and the possibility of quick updates for the FMS provider, who only has to add the updated component. As a result the DSS does not depend on a single user interface, but several interfaces will be built. Some of these will appeal sufficiently to users to make the DSS successful.

The **NemaMod** database, in which the parameters of the different models for each nematode/host combination are stored, is external. It can be updated or extended with new combinations of nematode and host without the necessity of changing the **NemaMod** component itself. This feature makes the engine independent of the current state of knowledge concerning parameter values and their stochasticity. New results can be incorporated in the component by updating the database. This not only extends the use of the component in applications, but also provides the possibility of adapting parameter sets for different countries, if necessary.

In Figure 2 some of the basic features of the engine are displayed.

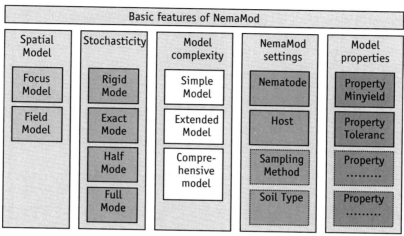

Figure 2. Basic properties of the NemaMod component.

- Spatial Model: There are two spatial models implemented in **NemaMod**. One is the 'Focus' model (Schomaker & Been, 1999) implying that the infestation detected is an infestation focus in an otherwise uninfected field. The other is the 'Field' model (Schomaker & Been, unpublished) which implies that the whole field is infested as is the case in growing areas with short rotations (1:2, 1:3).
- Stochasticity: **NemaMod** is based on a stochastic simulation model. This means that calculations are based on frequency distributions of parameters in the sub-models and uncertainties of estimations in the databases. The number of classes into which frequency distributions of parameters are divided can be chosen freely and so can be optimised for the computer system on which the DSS is used. Moreover, different stochastic settings are implemented for special use. For instance, when the ActiveX component is used as an educational tool, stochasticity can be suppressed completely using the 'Rigid Mode' setting when only average values are used. In 'Exact Mode' setting, the variables, e.g. the population density, are assumed to be known precisely and only the stochasticity of the parameters will be used. In 'Full Mode' setting, all parameters and variables have stochasticity.
- Model complexity: The models used differ in complexity from simple to comprehensive. E.g. for the relation between nematode density and yield this means that the simple model only addresses the mechanism of growth reduction at low to medium nematode densities. The extended model uses a compound model that

incorporates also the second mechanism of growth reduction, which becomes noticeable at medium to high nematode densities. The comprehensive model also includes a phenomenon called 'early senescence' which occurs at high nematode densities when certain cultivars are grown. That third model is added solely for educational purposes in the **NemaMod** component, as its practical use for advice is limited.

- Settings: Some major settings must be chosen before calculations can proceed. These include nematode species; crop and cultivar to be grown, the sampling method which has been used to estimate the initial nematode density, and the soil type.
- Properties: Every combination of nematode and host has its specific set of parameters - and their statistical properties - to be used in the different models. This information is available in the **NemaMod** database. Choosing a nematode/host combination automatically results in the relevant data-set being loaded. However, each parameter can also be set individually to values differing from that in the database. This functionality is provided for educational and exploratory use. Its implementation towards the user depends on the interface developer.

To date, the following methods have been developed for both the focus and the field models (table 1).
Demonstration interfaces for both advice and education aspects have been developed for both the focus and the field models in order to facilitate the development of the **NemaMod** and **SampView** component.

Focus model

Figure 3 displays a screenshot from the interface of the educational focus model. The left side of the screen represents the actual infestation, here potato cyst nematodes, reconstructed from soil sampling data. The farmer can choose to visualise the number of cysts, the number of eggs per g soil or the expected reduction in yield per square metre. He can select the nematode species and pathotype as well as the type of potato crop: seed-, consumption- or starch potatoes. Automatically, combo boxes with available susceptible, resistant and partial resistant potato varieties (Philips, 1984; Seinhorst, 1984; Seinhorst & Oostrom, 1984) will be filled. Partially resistant potato varieties are displayed with their relative susceptibility presented in percentages (Been & Schomaker, 1994). An information button (info) gives access to an information screen displaying all other characteristics of the chosen cultivar (see Figure 6).
The farmer now can choose various management measures, like growing a host or a non-host crop, or applying a nematicide (fumigant or non-fumigant) and will then see the simulated effect displayed both in the 3D graph and in the financial information frame. Moreover, farmers can calculate the probability of detection of the focus at any given time, using any of the known sampling methods used in The Netherlands. Farmers can investigate which crop measures will lead towards small cost/benefit ratio's and a low probability of detection by statutory soil sampling. The long-term impact of cropping

*Table 1. Methods of the **NemaMod** component and a short explanation of their action.*

Method name	Task	Used for PCN
NemaMod...		
ControlDataBase	Display edit/update window of **NemaMod** control database	yes
Damage	Calculate yield loss	yes
DataBase	Display edit/update window of **NemaMod** parameter database	yes
EnemyCrop	Calculate effect of a nematode trap crop on population density	yes
Evaluate	Calculate cost/benefit of control measure	yes
Fumigate	Calculate effect of fumigant application on population density	yes
FocusDetect	Calculate detection probability of focus	yes
GetBasicFreqs	Get text information on parameter frequency distribution	yes
GetFocusData	Get result of simulated cropping measure on infestation (focus)	yes
GetFieldData	Get result of simulated cropping measure on infestation (field)	yes
GetTextOutput	Get text info on simulation result	yes
Granulate	Calculate effect of a non-fumigant on population density	yes
GreenManure	Calculate effect of green fertilizer on population density	no
Host	Calculate effect of a (partially resistant) host	yes
MonteCarlo	Calculate prob. of detecting 0, 1, 2...xx cysts in a soil sample	yes
NonHost	Calculate effect of a non host on population density	yes
Redo	Redo last simulation step	yes
Reinitiate	Reinitialise the component	yes
WinterDecline	Calculate population decline during winter	no
Undo	Undo last simulation step	yes

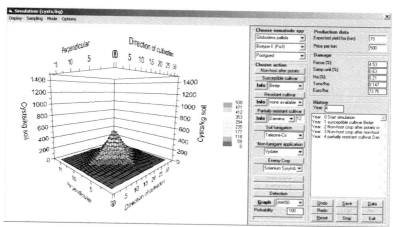

Figure 3. Screenshot from the interface of the educational focus model,, displaying actual infestation on the left, here in numbers of cysts, and actions and financial information on the right.

partially resistant cultivars on the population density of populations of *G. pallida* can be predicted. Also, long-term risks and financial losses caused by inadequate sampling methods can be calculated and visualised.

Field model

A screenshot from the interface of the field model for advisory services is displayed in Figure 4. The functionality of the interface is comparable to that of the focus interface in Figure 3. A graph displays the average values of loss of yield (bars) in the years when potatoes have been grown and the average population densities (line) throughout the years. However, when using the Field model and Full Mode stochasticity, the result of a calculation is a frequency distribution of possibilities and not just an average. This information, which is used for risk assessment, can be obtained by clicking on the x-axis label of the graph or selecting the year from the graph combo box (Figures 5A and 5B).

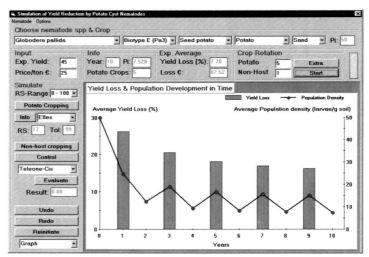

Figure 4. Demonstration interface for field infestations. Average values for loss of yield and population density are displayed in the main graph.

The initial population density is an estimate of the actual population density in the field. Depending upon the sampling method used to obtain this estimate, stochasticity is added. All current Dutch sampling methods have been evaluated using Sample IV (Been & Schomaker, 2001) and this information is incorporated in *NemaMod*. The frequency distribution of the initial population density is used as the primary input in the simulation. As all parameters of the sub-models have known frequency distributions, *NemaMod* can support stochasticity throughout all simulation steps. Therefore, for any given year the farmer can obtain a risk analysis for loss of yield. In Figure 5 the result

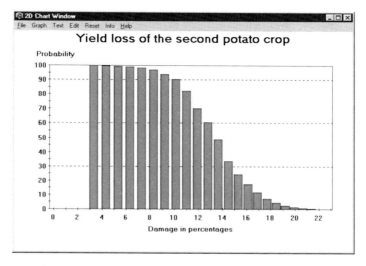

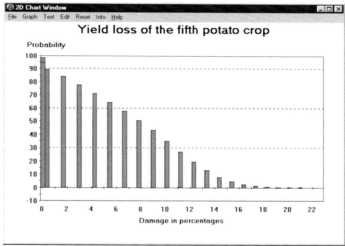

Figure 5. Screenshot of chart window presenting the expected risk of suffering a certain percentage of yield loss, expressed in % relative yield. A: yield loss of second potato crop. B: yield loss of fifth potato crop.

of one of these risk calculations for potato cyst nematodes in starch potatoes is given for two different years.

The cultivar Elles (relative susceptibility for Pa3 = 17%) is grown on an infestation with an initial average density of 50 larvae/g soil. Average loss of yield is high at 42.5%, but nematode densities after cropping are reduced. A 1-in-2 cropping rotation is simulated; the yield loss in the next potato crop is displayed in Figure 5A. The x-axis displays the minimum loss of yield as a percentage. The y-axis shows the probability of occurrence.

Although the average reduction in yield is 13.9%, losses are always greater than 3% of yield and never exceed 21%. Yield losses greater than 10% have a 90% probability. Figure 5B presents the risks of loss of yield when a fifth crop is grown. Here, the average loss is 8.2% and it will only decrease slightly over the following years if the rotation is continued with Elles or a cultivar with a similar relative susceptibility. A nematicide treatment must only be considered if pre-planting populations are so high that the risk of a calamity (defined by individual farmers) is high, say larger than 90%. The 'Evaluation' button in the Simulation frame automates the calculation of the cost/benefit ratios of control measures, such as fumigant and non-fumigant applications, trap crops and fallow. These calculations require a financial database, which is also linked to **NemaMod**.

The farmer can access an information window (Figure 6) displaying other quality aspects of potato cultivars, apart from resistance to and tolerance of nematodes. Farmers can make their own decision on the order of importance of desirable qualities in a potato cultivar. After selection of the cultivar the corresponding information - name, resistance and tolerance - will appear in the input fields in the main form.

Geo information and monitoring

Since 2002 the Dutch government has employed a Geo-database containing geo-data for all farmers' fields in the Netherlands. The Geo-data consists of the polygons of fields

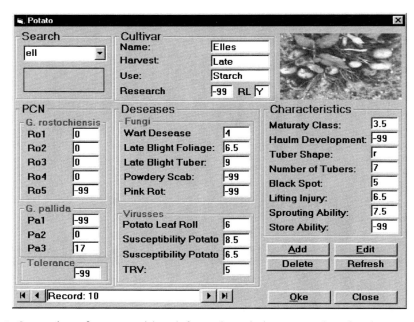

Figure 6. Screenshot of potato cultivar information window presenting the characteristics of the selected cultivar. Authorized users are allowed to modify the database.

and crop areas within the fields. Additional information such as aerial photographs are also available in the database. One of the ways to use this database is to visualise monitoring data related to cropping areas within fields. Figure 7 shows an example of a typical Dutch field sampled according to the AMI-50 scheme. This is an intensive sampling method for seed potatoes in a 1-in-3 rotation. The field is divided into strips with a width of 6 m and length of approximately 300 m in the direction of cultivation. *SampView,* which contains the sampling logic of soil sampling agencies and geometrical models and software, uses the field polygon provided by the Geo-database and any visualization component to display the monitoring data (Figure 7). Here two infestations are depicted, one with *Globodera pallida* (green) and the other with an unknown potato cyst nematode species (blue).

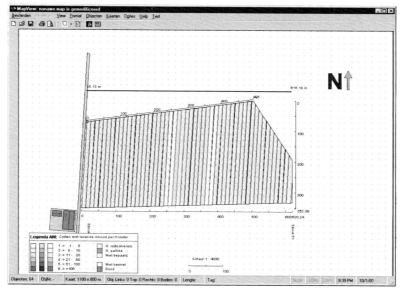

Figure 7. Using available Geo-information, the sampling logic, geometry software and sampling results a disease map for potato cyst nematodes is constructed.

Experience

NemaMod is an expanded version of a DOS programme developed between 1988 and 1992 for the potato industry. It was tested and validated in the starch potato growing area where there were huge problems with potato cyst nematodes. There, potatoes were grown in 1-in-2 or 1-in-3 crop rotations, always preceded by soil fumigation in the autumn before planting and application of a non-fumigant nematicide at planting. The program, named 'Seinhorst', was used between 1993 and 1997 as a scientific tool to investigate the possibility of reducing the use of nematicides in this area by the use of sampling schemes with known accuracy and potato varieties with partial resistance. In 1993 a large, multi-rotational, field-test was set up with 20 fields, each of 1 ha,. *G. pallida* was present in all the fields and, depending on the initial population densities of the nematode and the estimated loss of yield (according to the simulation) a decision was made whether or not a final, sanitary soil fumigation was needed before planting the new potato crop to give an acceptable initial population density. Non-fumigant nematicides were not applied. Cultivars were chosen for resistance to other soil borne diseases present. In 1995 or 1996, depending on the preferred cropping frequency, a second partially resistant cultivar was grown. No soil fumigation was conducted and nor were non-fumigant nematicides used in the second crop as population densities had been reduced sufficiently by the first partially-resistant cultivar and the decline in nematode population since that potato crop. It appeared that the population decline in the first year after a potato crop was much larger than reported in the literature.

Elles is the potato cultivar used most frequently in these field experiments although it is, relatively, not very susceptible - approximately 17% to the most virulent pathotype of *G. pallida* in the Netherlands - as it maintained population densities sufficiently high to be estimated accurately. Average potato yields in the field tests were 50 tonnes / ha and, after correction for the dry matter concentration, 60 tonnes / ha of payable weight. This yield was high, compared to a standard payable yield of 35 tonnes / ha for that area at that time. In Figure 8 the decrease in PCN population in all 20 test fields (F01-F20) is presented. At the start of the field test in 1993, population densities were extremely high in some of the fields, Higher, indeed, than had ever been reported after

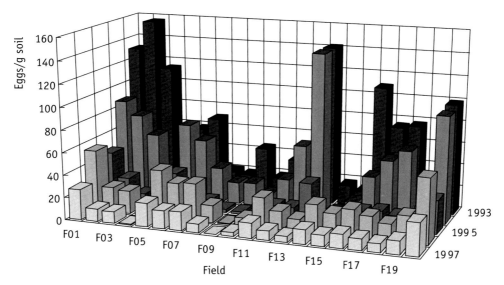

Figure 8. Decline of population densities in 20 fields of 1 ha, in the Dutch starch potato growing area before and after cropping partially-resistant cultivars in 1-in-2 and 1-in-3 crop rotations.

a non-host crop. That was because inaccurate monitoring systems, used previously, generally underestimate population densities. At the start of the third potato crop, densities of potato cyst nematodes had been decreased in almost all fields to economically acceptable levels of about 20 eggs/g soil. Lower densities of PCN and higher yields can be obtained by using varieties with higher degrees of partial resistance, which are becoming more abundantly available in The Netherlands. This is only feasible if nematode resistance is combined with other cultivar characteristics for pest resistance and quality.

Prospects

NemaMod will be expanded by adding relevant quantitative knowledge from different research programs instigated by government, trade, industry and branch organizations. *NemaMod* is the first step towards an integrated crop protection system. In the near future, *NemaMod* will be combined with modules for the control of fungal, bacterial and entomological diseases. This crop protection system will incorporate all available and relevant quantitative knowledge and be a tool to disseminate knowledge for education and practical use. The system will enable interest groups to pinpoint gaps in essential knowledge and to direct research in crop protection.

References

Been, T.H., C.H. Schomaker & J.W. Seinhorst, 1995. An advisory system for the management of potato cyst nematodes (*Globodera* spp). In: A.J. Haverkort & D.K.L. MacKerron (Eds). Current issues in production ecology, Dordrecht, The Netherlands, Kluwer Academic Publishers: 305-322.

Been, T.H. & C.H. Schomaker, 2000. Development and evaluation of sampling methods for fields with infestation foci of potato cyst nematode (*Globodera rostochiensis* and *G. pallida*). Phytopathology 90: 647-656

Phillips, M.S., 1984. The effect of initial population density on the reproduction of *Globodera pallida* on partially resistant potato clones derived from *Solanum vernei*. Nematologica 30: 57- 65.

Schomaker, C.H. & T.H. Been, 1999. A model for infestation foci of potato cyst nematodes *Globodera rostochiensis* and *G. pallida-*. Phytopathology 89: 583-590.

Seinhorst, J.W., 1984. Relation between population density of potato cyst nematodes and measured degrees of susceptibility (resistance) of resistant potato cultivars and between this density and cyst content in the new generation. Nematologica, 30: 66-76.

Seinhorst, J.W. & A. Oostrom, 1984. Comparison of multiplication rates of three pathotypes of potato cyst nematodes on various susceptible and resistant cultivars. Mededelingen Faculteit voor Landbouwwetenschappen, Rijksuniversiteit Gent, 49/2b: 605-611.

Name of the DSS

PLANT-Plus

Who owns the DSS, Principal author's, address,...

Dacom PLANT-Service BV, Waanderweg 68, PO Box 2243, 7801 CE
Emmen, the Netherlands

What is it for, what questions does it address?

Helping farmers to decide in their day to day operation on fungicide treatments, insect control and irrigation management.

Disease control: does the farmer have to spray and if so, what type of chemical should he best apply. Insect control: monitoring of insect population to decide on timing of chemicals. Irrigation management: continuous monitoring of soil moisture to determine refill point, best water gift and timing.

What input is required?

The system requires the availability of local weather data, either from synoptic or on-farm weather stations.

Who is it for, who are the intended users?

- Growers: to ease the day to day decisions
- Consultants: to assist in recommending farmers what to do
- Processors: to justify inputs

What are the advantages over conventional methods (whatever those may be)

Increased crop health and quality and lower inputs whenever the conditions allow to extend intervals.

How frequently should the DSS be used or its values be updated?

Effective use of the system requires daily communication with the databank to update the weather data and run the model calculations.

What scientific / technical enhancements would be desirable?

The models have proven to work in different countries and different climates.

11. PLANT-Plus: Turn-key solution for disease forecasting and irrigation management

P. Raatjes, J. Hadders, D. Martin and H. Hinds

PLANT-Plus was developed initially by Dacom as a DSS for management of *Phytophthora infestans*. It has been used on-farm since 1994. The system has been extended with a model for *Alternaria solani* and models for irrigation management based on ET data and soil moisture sensors. This way it offers a turn-key solution for the potato grower. The PLANT-Plus platform enables communication of data between farmer, consultant, processor and other accredited users in the food chain. The user can choose the most appropriate interface, such as Windows software (PC based) and Internet Server application and can configure a variety of output types such as SMS text messaging, Fax and Email warnings. PLANT-Plus offers an integrated five day weather forecast which provides a predictive risk assessment for the coming days.
The disease models require the availability of on-farm, automatic, weather data. All PLANT-Plus disease models are developed in cooperation with experts from several areas and countries, such as Dr. L.J. Turkensteen, Dr. H.T.A.M. Schepers, Dr. W.G. Flier and Phd. J.E. van der Waals. In contrast to most of the other available models, PLANT-Plus uses a biological model that is based on the lifecycle of the fungus and combines infection events with the unprotected part of the crop. The model will recommend when to apply a new spray and what type of chemical to use: contact, translaminar or systemic.
The benefits of the models are clearly demonstrated in field trials and commercial evaluations all over the world: PLANT-Plus is sound technology that provides safe spraying programmes with the lowest possible use of chemicals for the control of Late & Early Blight.
The models for irrigation management are based on ET_0/ET_c calculations of the soil water balance and/or direct monitoring through the use of soil moisture sensors. The ability to set the levels for field capacity and refill point, combined with the graphical outputs, allow the user to define when to irrigate and the most appropriate amount of water to apply.

Introduction

Dacom PLANT-Service BV is a commercial company from Emmen (NL) that has developed and operates the PLANT-Plus integrated system for crop management. Plantsystems Ltd is a UK based crop consultancy that has implemented the PLANT-Plus system across the UK, France and Portugal.

The development of the PLANT-Plus system was initiated in the early 90's by Dacom as a crop recording system. It was already then envisaged to provide a platform to be able to exchange data and information between various partners in the agribusiness.

On top of this platform a disease model was developed to optimise the control of *P. infestans* in potatoes. Given the state of technology at that moment it was possible to build a biological model based on fuzzy-logic principles and the use of hourly weather data, both retrospective and forecast. The PLANT-Plus databank uses a climate data interface that can import and distribute data of different types of on-farm weather stations, like Adcon, Skye, Metos, Campbell and Hardi Metpole. Holland Weather Services from Soest (NL, a WNI member) provides Dacom with a high-quality, local weather forecast for the coming 5 days for any desired location in the world.

The use of a databank platform also allows for the use of different user interfaces like old-fashioned PC-based MS-DOS applications, sophisticated PC-based MS-Windows applications and simple Internet Server Applications. The system can also deliver specific additional outputs like SMS text messages, E-mail and fax. The platform was subsequently extended with more regional applications: telephone-based, like ALPHI (Bouwman & Raatjes, 2000) and ALERT (Hadders, 2002) or like the web-based Syngenta Blight Forecaster (Hinds, 2003). The combination of crop records, weather data and treatment recommendations provides an excellent tool for crop assurance schemes.

Disease forecasting models & principles

The PLANT-Plus disease forecasting models have been developed in close cooperation and harmony with experts on plant diseases, like Dr. Turkensteen (NL), Dr. Schepers (NL), Dr. Flier (NL) and PhD. Van der Waals (SA). For potatoes the model for Late Blight (*Phytophthora infestans*) was started in 1994 (Hadders, 1997) and the model for Early Blight (*Alternaria solani*) was started in 1999 (Smith, 2000).

All model inputs and outputs are evaluated and calculated hourly or three-hourly for the forecast data. This implies the need for high quality continuous weather monitoring stations and regularly updated weather forecasts.

In contrast to most other systems the PLANT-Plus model is a biological model, that evaluates the complete life cycle of the fungus. Additionally, any infection events are related to the degree of protection of the crop by chemicals and to new growth of leaves. The model can be divided into the following sub-models:

1. Unprotected part of the crop
   a. Growth of new leaves
   b. Degradation and wear off of chemicals
2. Infection events of the disease
   a. Formation of spores on each infected leaf
   b. Ejection and dispersal of spores into the air
   c. Germination of spores and penetration into unprotected leaves
3. Combination of un-protected leaf area and infection events into treatment recommendations

Sub-model 1a: Unprotected crop by growth of new leaves
This factor is the most underestimated one in standard crop spraying regimes. Rapid growth in early season can cause crops to be vulnerable to infection from 3 days after fungicide treatment onward (Flier, unpublished). The assessment of growth of new leaves is dependant on field-scout reports, which means the farmer or his consultant will have to go out into the field regularly and score the development of new leaves compared to the last measurement(Hadders, 1997).

Sub-model 1b: Unprotected crop by degradation and wear-off of chemicals
PLANT-Plus includes information about chemicals as part of the base of the system. This information includes active ingredients, recommended dose rates, a.i. efficacy against the disease and factors for wear-off influenced by precipitation and solar radiation. The background data is mainly derived from independent trials that were recently carried out by Schepers and his colleagues at Plant Research in the Netherlands (1996-2002) and from the agrochemical companies.

Sub-model 1: Unprotected part of the crop
The results of sub models 1a and 1b together represent the unprotected leaf area. Basically it is of no concern if the crop is unprotected, as long as the there are no possibilities for the fungus to infect the crop. Trials have demonstrated that the spray interval can be stretched to 2-3 weeks without any problems under such circumstances (De Visser & Meijer, 2000; Raatjes *et al.*, 2001; Kessel *et al.*, 2003).

Sub-model 2: Infection events of the disease
This sub-model replays the life cycle of the pathogen in hourly steps. The development of 'new blight' with sexual reproduction and more aggressive strains of P. *infestans* (Flier *et al.*, 2002) is integrated into the system. The epidemiological soundness and accuracy of the Dacom models for *Phytophthora* and *Alternaria* have been successfully evaluated in numerous field trials (Smith, 2000; Denner & MacLeod, 1998; Marquinez, 1999; De Visser & Meijer, 2000; Wander & Spits, 2001). Whenever necessary the models will be updated to implement the most recent scientific background. The relation between a general fungus life cycle and the PLANT-Plus sub-models is presented in Figure 1.

Sub-model 2a: Formation of spores on each infected leaf
Sub-model 2a calculates the number of viable spores on an imaginary lesion using temperature and relative humidity ranges to simulate growing and dying. In effect it can be compared to the size of the grey ring of sporangiophores present around a developing P. *infestans* lesion. The source of the lesion can either be from infected seed, resulting in a primary infected shoot or from secondary leaf infections. For *Alternaria* it can also come from infected debris.

Sub-model 2b: Ejection and dispersal of spores into the air
After formation, the spores can be dispersed into the air. This can be caused by either climatic conditions like a (sudden) drop in the relative humidity, wind, or rain. But the

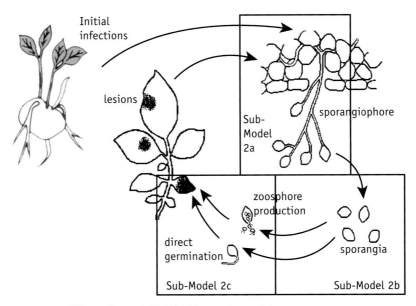

Figure 1. Fungus life cycle and PLANT-Plus sub-models.

lesion itself also purges out spores, a process called leakage (Turkensteen, 1995, unpublished). The inputs for this model are the output from sub-model 2a and the presence of the disease in the vicinity of the field, provided by field-scouts (Hinds, 2000) or a disease mapping system (Hendriks, 1999; Hadders, 2002). This sub-model has been evaluated with spore traps. The graph in Figure 2 compares the PLANT-Plus output with spore trap readings near a trial carried out by Schlenzig from TU München-Weihenstephan. Note that the absolute spore counts cannot be compared with the PLANT-Plus output as it calculates a fictitious figure, but the trends match closely.

The information about the presence of the disease is vital for an accurate calculation. The model will, nonetheless, calculate spore flights based on a very low (standard) presence, but not at accurate levels. Recent research has revealed the effect that different sources of infection can have on Late Blight epidemics, like infected mother tubers, dumps, (excessive) distant sources and volunteers (Flier *et al.*, 2002; Raatjes & Kessel, 2003; Van Baarlen & Raatjes, 2001). For example, in the Netherlands in 1999 one field with a severe infection caused secondary infections in other fields up to 30 km away. In 2000 and 2001 the effect of infected seed was clearly demonstrated and in 2002 early volunteers had great effect on primary inoculum levels. Sub-model 2b results in a fictitious concentration of airborne spores to be deposited on the foliage of the potato field.

Ad sub-model 2c: Germination and penetration of spores into unprotected leaves
The next step in the life cycle is to calculate the germination and infection of the spores on a leaf and it completes the infection event. This part is based on temperature, leaf

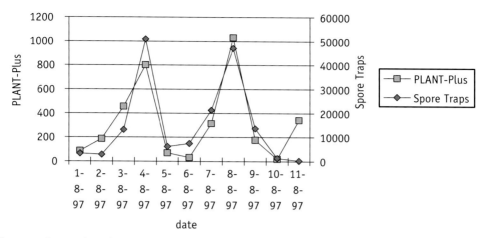

Figure 2. Comparison between output PLANT-Plus sub-model 2b and spore trap data.

wetness and variety resistance. Leaf wetness enables the spores to germinate. PLANT-Plus has a specific model that calculates the leaf wetness of the crop, based on climatic conditions and the latest observation for crop density. Temperature and variety resistance influence the speed of germination, penetration and incubation. Sub-model 2c results in the fictitious number of spores that can infect an unprotected leaf.

The accuracy of this sub-model is demonstrated in the correspondence with outbreaks in the fields (Van Baarlen & Raatjes, 2002). Based on surveys of Late Blight, age and size of lesions are associated with the timing of the infection events according to the model.

In a broader range the infection events for *P. infestans* were compared to the number of outbreaks reported to the disease mapping system in the Netherlands. The graph in Figure 3 shows the timing of the infection events compared to actual reported outbreaks of Late Blight. The arrows depict the delay in onset, caused by the incubation of the disease.

Raatjes & Kessel (2003) have also demonstrated that the accumulated PLANT-Plus infection index over the seasons reflects the accumulated number of outbreaks.

The outputs of models 1 and 2 are combined into one simple graph (Figure 4) that reveals all the necessary information. The model run always starts with the date of crop emergence or the most recent chemical treatment (left of the graph) and uses retrospective weather station data until the present (purple line) and continues with five days of forecast data.

Sub-model 3: Combination of unprotected leaf area and infection events into advice
Sub-model 3 interprets the unprotected leaf area and the infection event and provides a recommendation whether to use a chemical and what type (contact, translaminar or systemic) to use (Figure 5). The choice of the chemical is left up to the grower or his advisor. Within the recommendation, the system also specifies the relative need for a

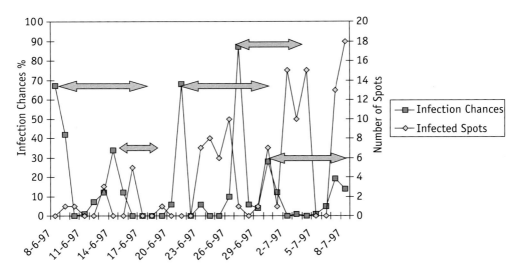

Figure 3. Relation between timing of PLANT-Plus infection events and reported outbreaks.

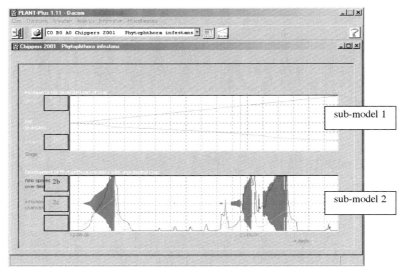

Figure 4. Example of PLANT-Plus disease model output.

treatment: not needed, to be considered or necessary. The following recommendations for chemical types are feasible:

- no treatment needed
- treatment with contact fungicide to be considered / necessary
- treatment with translaminar fungicide to be considered / necessary
- treatment with systemic fungicide to be considered / necessary

```
-- Calculated infections ------------------------------------------------------
      stage from              until              level
      D      12/06/00 12:00 - 19/06/00 18:00        33
      C      19/06/00 18:00 - 21/06/00 10:00         0
      B      21/06/00 10:00 - 22/06/00 20:00        10
      A      22/06/00 20:00 - 25/06/00 12:00       311

      advice threshold:  50 consideration 200 execution

== Advice =====================================================
```
| sub-model 3 |

```
      Application of Contact Fungicide necessary
-- Weather forecast -----------------------------------------------------------
                  Temperature       Rainfall   Radi-   RH      Wind
      Day   date    av.   max.  min. chance mm  ation  min. dir.   m/s
      fr 23/06/00   16.4  20.0  12.7   50   0.1  25.430  55  NW     4.9
      sa 24/06/00   16.2  19.8  11.5   60   0.6  27.220  50  W      3.8
      su 25/06/00   16.7  21.9   9.8   30   0.1  21.530  55  NW     3.6
      mo 26/06/00   16.0  21.8   9.0   20   0.0  24.850  45  WSW    4.2
      tu 27/06/00   15.0  17.3  12.4   40   3.6  16.615  60  W      5.2
-- Spraying conditions --------------------------------------------------------
               04 05 06 07 08 09 10 11 12 13 14 15 16 17 18 19 20 21 22 23
      fr 23-06                            -  -  -  -  -  -  O  -  -  O  +  +
      sa 24-06 +  +  +  O  ++ +  O  O  -  -  -  -  -  O  -  -  -  O  +  +
      su 25-06 ++ ++ ++ O  +  O  O  O  -  O  -  -  -  -  -  -  -  O  +  +
      mo 26-06 -  -- -- +  +  O  O  -  -  -  -  -  -  -  -  -  O  O  O
      tu 27-06 -- O  O  O  O  O  O  O  -- O  -  -  -  -- -- -  -  -  -  -
-------------------------------------------------------------------------------
```

Figure 5. Example recommendation output of Dacom disease forecasting model.

This recommendation is influenced by the timing of the infection event related to the current point of time. An infection in the previous 12 or next 24 hours will have to be treated with a contact fungicide. Depending on timing, temperature and variety resistance an 'older ' infection event will have to be 'cured' with either a translaminar or a systemic fungicide. The last three days of the forecast are not converted into an advice, but the user can view it in the graphical output.

Trials and projects have demonstrated that PLANT-Plus strategies rely heavily on contact fungicides as recommendations are often triggered before or during the infection event (Bouwman & Raatjes, 2000; Kleinhenz & Jorg, 2000, 2001). Continuous, daily consultation of the system will however be necessary to achieve this.

The PLANT-Plus model is not 'blocked' on short intervals or long intervals. This means that a new recommendation to treat can be given after 2-3 days, when conditions are dangerous and the crop is growing rapidly. On the other end recommendations not to spray can continue to be given for 3-4 weeks under dry conditions. As McGrath (2000) reports: farmers would normally never dare to wait that long.

All recommendations include an outlook for the spraying conditions for the next five days, based on expected rainfall and wind speed. This provides the user with a tool for advanced planning.

Principles for irrigation scheduling

The PLANT-Plus platform has been extended with models for irrigation management. The system offers two options to optimise the crop water usage:

Direct sensing based on soil moisture sensors. The climate data interface has the option to read data from two types of soil moisture sensors: suction and volumetric. The suction type sensors such as Gypsum block, Watermark and Tensiometer provide data in pF and hPa.

The C-Probe, Profile Probe and Virrib are volumetric sensors and those provide the relative soil water volume in %. The system has simple graphs to present, view and interpret the readings. Users can define custom settings for re-fill points and field capacity.

Calculation of Soil Moisture Water Balance based on ET_0.

The ET model in PLANT-Plus is based on the measurements of the weather stations combined with crop observations such as crop stage and crop growth. First the ET_0 is calculated according to the official FAO guidelines (Allen *et al.*, 1998). The outcome is modified into the ET_c, based on the recorded crop information. Using the date of planting the crop as the starting point, the continuous soil moisture water balance is calculated, while adding rain or irrigation and deducting the ET_c.

In practice the models are not only used to increase yield and reduce water usage, but also to prevent defects such as potato common scab.

Quality of weather forecasts

One of the key advantages in PLANT-Plus is the delivery of a local five-day weather forecast anywhere in the world. This implies a strong dependency on the forecast that must, therefore, be of high quality. Dacom is continuously evaluating the accuracy of the forecast data by comparison to the on-farm meteorological stations.

In 1997 a study was done in the Southern part of the Netherlands (Maastricht area) to evaluate the results of PLANT-Plus (Geelen, 1997). Relative humidity is a crucial factor in the epidemiology of fungal diseases. In the comparison the forecast underestimated the readings for the humidity by approx. 13%, which meant that the realised humidity was always higher. Raatjes *et al.* (2001) have made the same comparison for Egypt and find more or less the same results, but this has not resulted in (more) curative sprays or missed infection events. This means that for critical conditions the forecast will indicate the infection event in time, although the details of the forecast may not always be fully correct. The conclusion is supported by the large share of contact fungicides that PLANT-Plus users generally apply.

Experiences and results

The experiences and results with PLANT-Plus can be split into two parts: field trials and commercial evaluations amongst growers and grower groups.

Field trials with P. infestans model

Many field trials around the world have been carried out to demonstrate the benefits of the PLANT-Plus *P. infestans* model. Some of the trials will be discussed here and the results are summarised in Table 1. The trials represent different climatic conditions: from dry, arid to wet, maritime climates. The density of potato crops in the area also differs from trial to trial.

In the Netherlands evaluation trials were conducted at research station 't Kompas in Valthermond over the years 1995 to 2001 where the PLANT-Plus model was compared with standard field practice. Using the PLANT-Plus model resulted in an average reduction of 28% in the number of applications, while disease control remained adequate.

In South Africa a trial was conducted in 1998 to compare different disease forecasting models with standard practices (Denner & MacLeod, 1998). This trial resulted in failure of the Winstel and Ullrich-Schrodter models to control *P. infestans* effectively under these high pressure conditions. The PLANT-Plus model managed to control the disease as well as the standard field spray programme, but reduced the number of sprays applied. It should be mentioned that most of the field trials do not represent 'normal' field conditions, but are under extreme pressure from the untreated, infected controls.

A trial in the UK at Harper Adams University College with the variety Cara in 1998 (Jenkinson, 1999) resulted in a very robust PLANT-Plus programme that was very cost effective, despite the fact that more sprays were applied compared with standard practice.

A trial by the Bundesamt für Forschung und Landwirtschaft in Austria resulted in no reduction in chemical use, but significantly improved disease control. In Spain in 1998 the summer was very dry and PLANT-Plus managed to keep fields virtually blight free with only 2 contact sprays, compared to 2 systemics and 2 contacts in normal practice (Marquinez, 2000). In Tulelake valley in the USA a trial was conducted in 1998 by UC Davis that resulted in improved control of the disease and a small reduction in the application of fungicides.

In the Netherlands a broader testing programme was started in 1999 to compare different advice models with standard field practice. The PLANT-Plus model resulted in a fair reduction in chemical costs of 60 Euros per hectare while maintaining appropriate disease control (Kessel *et al.*, 2003). Wander & Spits (2001) concluded after the first year that PLANT-Plus uses relatively few curative products and starts relatively late with the first spray, compared to other models, like ProPhy and NegFry. Kleinhenz & Jorg (2001 & 2002) concluded the same for trials in other countries such as Belgium, Switzerland, Ireland, and Germany. Generally the PLANT-Plus model performed well in the trials that were carried out within the framework of the EUNET project (Hansen,

Table 1. Results of field trials with PLANT-Plus P. infestans model This table compares number of sprays per season, chemical costs in Euros per hectare, foliage infection at the end of the season, tuber blight after harvest and yield in tonnes/ha for two strategies: STD (common practice in the area) and PP (PLANT-Plus).

Trials	Sprays (no.)		Costs (Euros/ha)		Foliar Blight (%)		Tuber blight (%)		Yield (tonnes/ha)	
	Std	PP	Std	PP	Std	PP	Std	PP	Std	PP
Netherlands,Kompas 1995-01, unpublished)	14,3	10,6	267	212	-	-	-	-	-	-
South Africa, 1998, Roodeplaat, Denner & MacLeod, 1998)	7,0	5,0	-	-	76,5	78,5	3,8	2,3	33,2	40,4
AustriaTrials at BFL 1998; unpublished)	7,0	7,0	-	-	69,0	42,0	-	-	-	-
Spain, Trial in 1998; Marquinez, 2000)	4,0	2,0	-	-	0,00	0,01	-	-	-	-
United States, 1998 Tulelake valley, U.C. Davis; unpublished)	4,0	3,0	-	-	10,0	5,0	-	-	-	-
UK, BPC, 1998; Jenkinson, 1999)	12,0	15,0	252	295	1,50	3,50	-	-	54,1	60,8
Netherlands, 3 trials 99-01; Kessel et al., 2003)	14,9	11,8	290	229	0,04	0,15	1,6	1,2	71,6	71,1
Sweden, 3 trials 1999-2001; Wiik, 2002)	10,9	9,5	-	-	0,02	0,05	0,6	2,8	-	-
Belgium, Gembloux, Hansen et al., 2001)	11,0	7,0	-	-	3,5	4,0	-	-	-	-
Ireland, Oakpark (Hansen et al., 2001)	13,0	11,0	-	-	50,0	15,0	0,15	0,03	62,2	72,4
Germany, 2001& 2002; LWK Viersen, unpublished)	10,5	11,0	430	437	59,0 (2002)	42,0 (2002)	-	-	73,8 (2002)	75,6 (2002)
USA, Tappen, 2002; Syngenta/NDSU; unpublished)	10,0	7,0	-	-	0,0	0,0	-	-	-	-

2002; Kleinhenz & Jorg, 2000,2001). PLANT-Plus provides an excellent and robust combination of reducing the number of sprays, while controlling the disease.

Trials by Wiik in the Southern part of Sweden (Skåne) from 1999 to 2001 have demonstrated that PLANT-Plus gives good control for starch varieties with nearly 30% reduction in the number of treatments. In the susceptible Bintje the model had some problems with recommendations in the first two years of the projectin one of the trial locations.

In 2002 a trial using PLANT-Plus was carried out by the LandWirtschaftsKammer Vierssen in Germany. It was estimated that the use of PLANT-Plus resulted in a financial revenue of approx. 200 Euros per hectare when yield increase and chemical costs were taken into account.

Commercial evaluation of P. infestans model

Dacom has always closely monitored the performance of the P. *infestans* model in commercial fields, either through projects or study groups. At present over 1.000 farmers all over the world are using this model. Table 2 summarises the results of some evaluations.

The results of the system have been studied ever since the introduction of PLANT-Plus in the Netherlands. In the North Eastern part of the Netherlands the STER project for sustainable agriculture (Hadders, 1997) demonstrated that using the PLANT-Plus model resulted in substantial savings on blight control (Regeer, 1997). The same results were found in other projects in other parts of the Netherlands in the years 1995 to 1999. The costs of both blight control and crop protection are greatly affected by the weather conditions during the seasons: 1995 and 1996 were rather dry and 1997, 1998 and 1999 were rather wet. At present it is rather difficult to compare the results of PLANT-Plus users with non-users as all potato growers in the Netherlands are alerted by telephone when PLANT-Plus calculates an infection event in their area (Hadders, 2002).

In an inquiry amongst PLANT-Plus users in Flevoland, Netherlands it was found that the confidence of growers was rated at 3,9 on a 1-5 scale (Balk, 2000). PLANT-Plus users applied on average more sprays than common in the area, but spraying costs were lower. The costs strongly correlated with the confidence of the user in the system.

Hinds (2000) studied the use of the PLANT-Plus model compared with farmer's normal practice in the UK and found a significant reduction in fungicide use, while there was no difference in blight control under moderate blight conditions.

In Egypt, winter conditions are relatively dry, but sometimes a devastating late blight epidemic can occur. Farmers tend to (over) spray with prophylactic sprays. The introduction of PLANT-Plus has resulted in approx. 50% reduction while keeping the crops disease-free. The fields of the project were under a centre pivot irrigation regime, but still the intervals could be stretched to over 3 weeks without any problem (Raatjes *et al.*, 2001). The feasibility of the model was also tested under tropical conditions in Indonesia which is a region where previously farmers have been unable to keep the crops alive until tubers are full grown. Implementation of PLANT-Plus resulted in postponing

Table 2. Results of commercial evaluations of P. infestans model. This table compares number of sprays per season, chemical costs in Euros per hectare and control of foliar blight at the end of the season (+++ very good; ++ good; + acceptable; - bad; - - very bad) for two strategies: STD (common practice in the area) and PP (PLANT-Plus).

Area	Year(s)	Sprays (no.)		Costs (Euros/ha)		Blight control	
		Std	PP	Std	PP	Std	PP
Netherlands (Project Westerwolde area)	1995	12,6	10,1	188	142	++	++
Netherlands (Study groups in STER-project; Regeer, 1997)	1996	10,4	9,0	154	136	++	++
Netherlands (Project Brabant Province)	1997	16,3	15,5	315	304	+	+
Netherlands (Project Drenthe Province)	1996-1999	13,5	11,7	217	197	++	++
Netherlands (Project Maastricht area)	1997-1998	14,8	13,1	305	253	+	++
United Kingdom (Commercial evaluation in the East Midlands; Hinds, 2000)	1999	12,2	10,0	243	176	+++	+++
Egypt (Feasibility study in Ismalia; Raatjes et al., 2001)	2000-2001	7,5	3,5	272	127	++	++
Indonesia (Feasibility study on Sumatra)	2002	14,3	10,0	441	241	—	++
Canada (On Farm trial on PEI (Canada); Department of Agriculture)	2002	9,0	6,0	-	-	+++	+++
Belgium (Farm Frites/ Syngenta project)	2002	13,5	11,3	295	219	+	++
Latvia (Syngenta/ Latfood project)	2002	5,0	4,0	-	-	+	++

the onset of the disease while saving the use of chemicals by increasing chemical efficacy through improved timing of applications. Recent evaluations in Belgium, Canada and Latvia have confirmed the previous results.

Hinds (2001) investigated the practicalities that large farms run into when they want to apply the PLANT-Plus system on large areas of potato and found it possible to implement the system. Reaction time to spray warnings can be within 2 days, given good communication and adequate sprayer capacity. A Syngenta/Frito-Lay project using the variety Courlan (a sensitive variety) emphasised that Late Blight can be effectively prevented by optimised timing of fungicide treatments according the PLANT-Plus model (Hinds, 2002), whereas the standard strategy might fail.

One of the other advantages of the PLANT-Plus model is that it can produce retrospective reviews of the season that can be used to teach and train the farmers about how to improve their late blight management strategy. Mistakes can be pinpointed and alternative strategies preventing late blight epidemics are provided.

From both the trials and commercial on-farm evaluations it can be concluded that PLANT-Plus is proven, sound technology that provides safe spraying programmes with the lowest possible use of chemicals for the control of Late Blight.

Choice of fungicide planned to control *P. infestans* has, in some cases, resulted in increased problems with the control of Early Blight (*A. solani*). This might be due to poor control of Early Blight by some oomycete-specific fungicides.

Field trials and commercial evaluations of the A. solani model

The model for *Alternaria solani* was first used as a prototype in cooperation with the University of Pretoria, South Africa by PhD. J.E. Smith-van der Waals. The model was evaluated and calibrated in field trials in 1999 and 2000 and the calculation of infection events by PLANT-Plus matched the actual spore flights monitored using spore traps (Smith, 2000). Based on the results of the trials the model was modified to increase accuracy.

Hadders (2003) concluded that infection events for *Alternaria* often coincide with the infection events for *Phytophthora* and that the problems with *Alternaria* are therefore not recognized because of the (side-)effect that the *Phytophthora* treatments have on *Alternaria*. On several occasions, however, separate infection events are observed for *Alternaria*.

In a project in Egypt aiming to study the technical and economical feasibility of the PLANT-Plus models it was recognized that successfully reducing the applications against Late Blight increased the problems with Early Blight. In the second year of the project the Early Blight model was therefore included in the study. A comparison at Chipsy's farm led to the conclusion that both standard and PLANT-Plus strategies provided adequate control of the disease, but PLANT-Plus applied six sprays, compared with eight in normal practice.

In a Syngenta trial in Tappen, North-Dakota the *A. solani* model was tested by Dr. N. Gudmestad (North Dakota State University) in the 2002 season. PLANT-Plus suggested seven sprays compared with ten following the standard schedule and ten in a Blitecast strategy. The incidence of the disease under the PLANT-Plus strategy was somewhat higher than under the other two, but we argue that this was largely influenced by the plot layout and by the choice of fungicide. It was decided in advance to use only Bravo Zn to control *Alternaria*, whereas a strategy with a combination of other active ingredients probably would have been more effective.

Field trials and commercial evaluations of the irrigation management model

The ET model for irrigation management in PLANT-Plus was validated in a trial at Kompas, Valthermond, Netherlands in 1997 and it resulted in 78% crop cover at the end of season compared to 74% in the unirrigated control. The Valthermond area with typical sandy/peaty soils is not an area where irrigation is usually needed.

In Japan, a comparison was made in 2001 & 2002 for Calbee Potato between the C-Probe, Watermark and the ET model, based on 10 locations on Hokkaido. It was concluded that the calculations from the ET model matched the trends from the Watermark and C-Probe sensor readings (Figure 6) and can therefore be used as a tool for irrigation planning.

Murata (Calbee, 2002, unpublished) validated the readings from the C-probes by oven-drying soil samples to analyse their water content and found a good correlation between the readings and the actual water content.

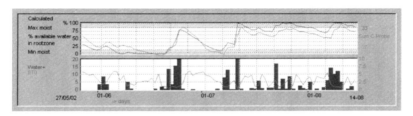

Figure 6. Comparison between summed C-Probe readings (red line) and calculated soil water balance (green line), based on FAO ET model.

In Portugal the use of tensiometers clearly indicated a failure in irrigation practice. Not being alert resulted in soil conditions that were too dry for only two days at around tuber initiation and the effect was a significant incidence of potato common scab.

The use of Watermark sensors with PLANT-Plus in a cotton crop in Egypt resulted in a saving of 300 mm of water applied by sprinkler irrigation compared with traditional furrow irrigation (Banna, 2002, unpublished).

The interpretation of the irrigation management models can be compared to trend-watching and is therefore rather complex and takes a lot of learning. A number of mistakes, such as incorrect sensor installation or non-representative positioning in the field, can have a large influence on the success or otherwise of the system. Common sense combined with technical skill will therefore always be needed when applying sensor- or ET-based irrigation strategies.

Future developments and constraints

The previous results and experiences might lead to the conclusion that every farmer should be ready to use the PLANT-Plus system as there is a clear economic benefit. Yet, even in the Netherlands there is only a small, although steadily increasing, proportion of the 'professional' farmers using the models. Spanninga (1998) concluded in his market survey that farmers have different kinds of excuses why not to use the models. One of them is the complexity of the subject and the difficulties farmers find in understanding the recommendations. To change the farmer's mind from an easy-going, prophylactic strategy to an alert approach, reactive to infection events has always been the toughest challenge. It is as a South-African vine grower said: "PLANT-Plus is like a cell-phone. First you don't want it, but when you have it, you can't be without it anymore."

Contradictory recommendations or even opposition from others who influence opinion, like chemical resellers or governmental bodies, is another threat to successful implementation.

Furthermore, the successful introduction of PLANT-Plus requires the availability of an effective range of fungicides for the control of the disease. In some countries like Sweden and the Netherlands there are only a few fungicides registered and this sometimes leaves the farmer with few options to choose products, especially translaminar and systemic fungicides. It should be a general concern that most of the registered chemicals are not fully efficacious against A. *solani*

Another point of concern is the effort that farmers have to make to provide crop information. In essence this is a positive thing as farmers will have to go out into the field to observe their crops, but it is often considered to be too much trouble. Dacom is therefore seeking alternatives such as Remote Sensing images or automated snapshot cameras. Bakker (1999) has found that processing RS images into crop cover maps can provide accurate information. The PLANT-Plus feasibility study in Egypt (Raatjes *et al.*, 2001) however, revealed significant problems with the availability of useful, cloud free images.

Dacom has extended the PLANT-Plus system with a large and successful range of disease models for other crops like vegetables, fruits and vines. Dacom is also developing additional models for insect management. Currently models for thrips, carrot fly and aphids are in the testing phase.

References

Special Report in the references refers to the annual Proceedings of the Workshop on the European network for development of an integrated control strategy of potato late blight, edited by H.T.A.M. Schepers.

Allen, R.G., L.S. Pereira, D. Raes & M. Smith, 1998. Crop evapotranspiration guidelines for computing crop water requirements, FAO Irrigation and drainage paper no. 56.

Baarlen, P. van & P. Raatjes, 2002. Karakterisering van primaire haarden van *Phytophthora infestans* tijdens het teeltseizoen 2001, project report Masterplan Phytophthora.

Balk, I., 2000. Het vertrouwen in een waarschuwingssysteem tegen Phytophthora infestans, Promotion Study at CAH Dronten.

Bakker, J., 1999. PHYTOGIS; Integrated service of using earth observation in optimization of services in Phytophthora disease control, Project report ITT RGC 11/99.

Bouwman, J.J. & P. Raatjes, 2000. ALPHI Actual Local Phytophthora Information Line based on PLANT-Plus, in PAV Special Report No. 6, pp 91-95.

Denner, F. & A. McLeod, 1998. Evaluation of disease forecasting models for control of potato late blight, Report ARC Roodeplaat, South Africa.

Flier, W.G., G.J.T. Kessel, G.B.M. van den Bosch & L.J. Turkensteen, 2002. Impact of new populations of *Phytophthora infestans* on integrated late blight management, in Special Report No. 8, pp 193-202.

Geelen, P., 1997. Het gebruik van het Dacom adviesmodel onder Zuid-Limburgse omstandigheden in 1997, Trial report for LLTB.

Hansen, J.G. *et al.*, 2002. Results of validation trials of Phytophthora DSSs in Europe, 2001, in Special Report No. 8, pp 231-242.

Hadders, J., 1997. Experience with a Late Blight DSS (PLANT-Plus) in starch potato area of the Netherlands in 1995 and 1996, in PAV Special Report No. 1, pp 117-122.

Hadders, J., 2002. Don't call us, we call you, in Special Report No. 8, pp 93-102.

Hendriks, H., 1999. Waarnemen en registreren seizoen 1999, project report Masterplan Phytophthora.

Hadders, J., 2003. Control of Alternaria with PLANT-Plus, in Special Report No. 9, pp 95-104.

Hinds, H., 2000. Using disease forecasting to reduce fungicide input for potato blight in the UK, in PAV Special Report No. 6, pp 83-90.

Hinds, H., 2001. Can blight forecasting work on large potato farms?, in PAV Special Report No. 7, pp 99-106.

Hinds, H., 2002. A late blight forecasting project for FL1953 (Courlan) a blight sensitive processing cultivar, in Special Report No. 8, pp 131-136.

Hinds, H., 2003. Blight Forecaster - a web based DSS using UK local and forecast weather data, in Special Report No. 9, pp 211-216.

Jenkinson, M., 1999. Evaluation of two decision support systems for the control of late blight in potatoes, Report Nr. 98/216, Harper Adams University College.

Kessel, G.J.T., J. Wander & H. Spits, 2003. Evaluatie van beslissingsondersteunende systemen voor bestrijding van *Phytophthora infestans* in aardappel, Project Report Agrobiokon.

Kleinhenz, B. & E. Jorg, 2000. Results of validation trials of Phytophthora DSS in Europe in 1999, in PAV Special Report No. 6, pp 180-190.

Kleinhenz, B. and E. Jorg, 2001. Results of validation trials of Phytophthora DSS in Europe in 2000, in PAV Special Report No. 7, pp 23-37.

Marquinez, R., 1999. Blight forecast and chemical control in Spain, experience in 1998, in PAV Special Report No. 5, pp 164-171.

McGrath, P., 2000. Beating blight saving sprays, The Grower, April 6.

Raatjes, P., H. Germs, M. Verbeek & M. van Loon, 2001. Phytophthora disease forecasting in Egypt, Project report PESP 00066,.

Raatjes, P. & G.J.T. Kessel, 2003. Phytophtora hotspots in de Veenkoloniën, Project Report Agrobiokon.

Regeer, H., 1997. STER-Project Duurzame Akkerbouw Resultaten 1996, project report.

Smith, J.E., 2000. Fighting early blight with PLANT-Plus, Chips Magazine Sep/Oct.

Spanninga, M., 1998. Aspecten die al dan niet leiden tot aankoop van het managementprogramma PLANT-Plus, Wageningen University Study Report.

Visser, C.de & R. Meijer, 2000. Field evaluation of four decision support systems for potato late blight in the Netherlands in 2000, in PAV Special Report No. 6, pp 137-155.

Wander, J. & H. Spits, 2001. Field evaluation of four decision support systems for potato late blight in the Netherlands in 2000, in PAV Special Report No. 7, pp 77-90.

Wiik, L., 2002. Bekämpningsstrategier mot potatisbladmögel, Slutreport SLU trials.

Name of the DSS

MLHD online

Who owns the DSS, Principal author's, address,...

Plant Research International b.v.

Wageningen University and Research Centre

P.O. Box 16, 6700 AA Wageningen, The Netherlands

Contact person: C. Kempenaar

corne.kempenaar@wur.nl

MLHD online is developed in cooperation with Opticrop b.v. in Vijfhuizen, NL.

What questions does it address?

MLHD online is a DSS for rational use of herbicides in arable and vegetable crops. It generates dose advice for photosynthesis-inhibiting herbicides based on information on the herbicide used, weed species and stages present in the crop, crop conditions, weather conditions and readings from sensing techniques. The DSS can be used in different ways, to minimise the risk of incomplete weed control while apply reduced herbicide doses, or to visualise and minimise the effect of herbicides on the crop.

What input is needed?

Data on weed species and stages present in the crop are the minimum requirements for using MLHD online. Other important inputs are measurements of herbicide efficacy and crop response taken by portable sensing techniques, weather data and crop development stage.

Who is it for, who are the intended users?

This decision support system is intended for:

- farmers who decide on herbicide doses
- extension specialists
- research specialists

What are the advantages over conventional methods (whatever those may be)

The advantages are:

- a further rationalization of herbicide use
- a reduction of herbicide use, costs and side effects
- insight to crop response and a tendency towards higher yields

---

How frequently should the DSS be used or its values be updated?

MLHD online should be used during the period when weeds are to be controlled in arable and vegetable crops.

What are the major limitations?

The most important limitation is that DSR are not yet available for other than photosynthesis-inhibiting herbicides, although these herbicides are also important. There is a current R&D programme to solve this matter, but it will take several years.

What scientific / technical enhancements would be desirable?

To extend MLHD online to other than photosynthesis-inhibiting herbicides and link the DSS to a DSS that takes into account weather effects on herbicide efficacy (*e.g.* GEWIS)

12. MLHD, a decision support system for rational use of herbicides: developments in potatoes

C. Kempenaar and R. van den Boogaard

MLHD, short for Minimum Lethal Herbicide Dose, is a Dutch concept that further rationalises herbicide use in integrated weed control. The concept has been developed since the mid 1990's, and comprises two distinct elements: (1) a weed species and stages dependent dose advice and (2) sensing techniques that allow predictions of herbicide efficacy and the need for additional treatments. The second element is an innovation in weed control; herbicide doses are partly based on readings from sensing techniques. The MLHD concept is ready for use in practice for photosynthesis-inhibiting herbicides, which are used in many arable and vegetable crops.
In this paper, the MLHD concept and background is explained. Results from studies focused on use of the MLHD concept in weed control and haulm killing in potatoes are presented. Examples of on farm research and farmers' perceptions are given. In addition, MLHD online, the web based MLHD decision support system and some prospects are presented and discussed.

Introduction

MLHD is a concept that supports rational herbicide use in integrated weed control. MLHD means Minimum Lethal Herbicide Dose. The concept was originally developed by Ketel *et al.* in the 1990's (*e.g.* Ketel & Lotz, 1997). Ketel developed decision support rules (DSR) which related a minimum lethal dose of photosynthesis-inhibiting herbicides to the size of annual weeds and he added a sensing technique to the concept that allowed an assessment of the efficacy of the herbicide treatment shortly after application. The latter is an innovation in weed management. The assessment is used to predict whether the reduced herbicide dose will be effective or if an additional treatment is needed.
The MLHD concept was tested and further developed in on-farm research from 1998 to 2002. The original MLHD concept was modified during that period using feed back of farmers and others (advisors, contractors, weed control scientists) involved. Today, the MLHD concept is a DSS for more rational use of photosynthesis-inhibiting herbicides. Advice on doses is specific to the given weed situation, with differentiation for weed species present, development stages of the weeds and assessments of herbicide efficacy. Sensing techniques that are used to assess the effects of herbicide treatments are absorption (PS1-meter), fluorescence (PPM-meter) and reflection (Crop scan meter). Weather effects are taken into account by simple comment lines.

Figure 1. MLHD-PS1-meter.

Since 2001, the MLHD DSS has been available on the internet in a model named MLHD online. In this paper, information on MLHD online version 2002 is presented. But first, the MLHD concept is explained in more detail for potatoes.

The MLHD concept

MLHD is a new concept that further rationalises weed control in arable crops. It fits in any weed control strategy where herbicides are used; in one-spray systems, in low-dosage systems (LDS) and in integrated weed control systems. The MLHD concept basically consists of three steps:
1. Advice on a minimum lethal herbicide dose is obtained from MLHD online (www.mlhd.nl), which generates advice on the basis of field-specific information on the crop, weeds (species and stages) and weather. Underlying decision support rules are explained in the next paragraph.
2. Shortly after application (2 days or more), a sensing technique is used to assess whether the dose will be effective. Measurements are done on 5 to 10 plants of the 5 most troublesome annual weeds in the crop (the readings will tell if the weeds will die or not). Sometimes measurements are also done on crop plants to assess whether the crop is affected by the treatment or not.
3. A decision is made whether additional weed control is needed based on the readings of the measurements under step 2 and crop and weed information. If readings are below a certain level, MLHD online advises an additional treatment. In an LDS, the measurements can be repeated shortly before the next treatment to improve the quality of the advice.

More information on the steps in MLHD is given in a manual in Dutch on the internet (Kempenaar, 2002). The aforementioned step 2 is only relevant for herbicides that affect photosynthesis shortly after treatment, but where it takes much longer before effects can be seen with the naked eye, which is *e.g.* the case for photosynthesis-inhibiting herbicides.

How does the MLHD measurement work? The PS1-meter, for example, produces a reading (figure) on a scale of 0 to 100. The leaf of a weed or crop plant is clipped in the PS1-meterwhich measures the proportional damage to the photosynthetic apparatus. The higher the reading, the greater the effect (damage) on the plant. In the case of photosynthesis-inhibiting herbicides, a reading of > 80 from the youngest measurable leaf two days after treatment predicts that the weed will die. A reading of > 65 indicates a good level of control, but some plants may survive. An example of predictions of herbicide efficacy based on such readings is given in Table 1. Details on the PS1-sensing technique are given in (Kempenaar & van den Boogaard, 2002). Figure 1 shows a picture of the PS1 sensing technique. The PS1-technique is developed and patented by ATO b.v. in Wageningen.

Table 1. Example of MLHD-measurements on young Solanum nigrum plants in a greenhouse experiment.

Dose of Sencor WG (80 % metribuzin)	Reading PS1-meter, 2 days after treatment	Efficacy 14 days after treatment
0 g/ha	9	No effect
10 g/ha	45	Growth reduction
50 g/ha	75	Dead plants
100 g/ha	85	Dead plants

The MLHD model

In the mid 1990s Ketel studied the relation between plant size and the minimum number of herbicide molecules needed to kill a plant (Ketel & Lotz, 1997, Ketel & Lotz, 1998). His studies were mainly done with photosynthesis-inhibiting herbicides metribuzin and fenmedifam and the weeds *Chenopodium album* and *Polygonum persicaria* (annual broadleaved weeds). From his studies, he derived the original MLHD formula:

$$MLHD = FW_{leaf} \times T \times MW_{herb} \times R \times 10^{-4}$$

Where:
FW_{leaf} = mean fresh weight of the leaves per surface unit [g]
T = target size, composed of the number of target sites (chloroplasts) per gramme of leaf fresh weight and the number of herbicide molecules needed to inactivate one target MW_{herb} is the molecular weight of the herbicide
R is a ratio parameter
and 10^{-4} is a scale constant.

Standard values for T and R were estimated with regression analysis on data from greenhouse experiments.

In the current MLHD model, Ketel's formula is the backbone for the advice on doses. However, some modifications have been applied since 1999. Fresh weight is translated into development stage because farmers are not interested in assessing the weights of weeds. Also, a parameter for weed sensitivity class has been introduced. Finally, a parameter that corrects for weather effects and formulation has been added. In the current version of MLHD online, corrections for weather effects are given in text boxes.

The PS1 sensing technique

The effectiveness of herbicides can be determined using the PS1 technique which measures light-induced changes in absorbanceat around 820 nm (Genty and Harbinson 1996). This wavelength is specific for reactions occurring in photosystem I (PS1). If herbicides block electron transport from photosystem II to photosystem I, reactions in photosystem I will be affected.

In the light, $P700^0$ is oxidised to $P700^+$, which is subsequently reduced to $P700^0$ once more by electrons coming from photosystem II. At 820 nm $P700^0$ and $P700^+$ have a different absorbance. In the PS1-meter a leaf is first placed in the dark, then a flash of light is applied. The oxidation and reduction following this flash of light is measured from the changes in absorbance at 820 nm. This gives a measurement of the extent of blocking of reduction of $P700^+$ by herbicides.

In Figure 2 the measurement principle is shown. In figure A there is no herbicide effect and after a pulse of light there is complete reduction of $P700^+$ to $P700^0$. The meter will indicate 0% damage. In figure B a herbicide blocks photosynthesis completely, $P700^+$ is not reduced to $P700^0$. The meter will indicate 100% damage. In case C there is partial inhibition of photosynthesis. In this case the meter indicates 45% damage.

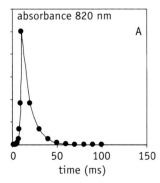 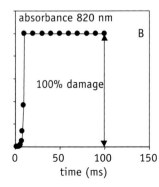 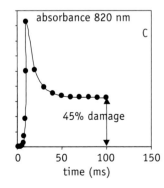

Figure 2. Measurement of absorbance at 820 nm indicates damage to the photosynthetic apparatus: A = normal functioning photosynthetic apparatus, B = 100 % blocked photosynthetic apparatus and C = 45 % blocked photosynthetic apparatus.

The dynamics of PS1-readings with time are shown in Figure 3. Mean values of PS1-readings at 1, 2 4 and 6 days after application of Sencor to young *Solanum nigrum* plants are shown. In the situation where the plants where lethally treated (180 g/ha), the PS1-values rapidly increase above 80. Sub-lethally treated plants show an increase in PS1-reading, but after some days the PS1-readings drop back to the levels of the untreated situation.

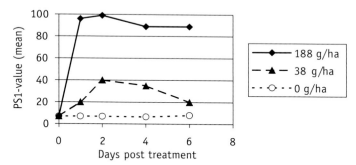

Figure 3. *PS1-readings on young* Solanum nigrum *plants treated with differing doses of Sencor in a greenhouse pot experiment.*

On farm research

Since 1998, MLHD has been tested in on farm research. The testing was done on commercial arable crops on farms in the Netherlands. In the tests, one part of the crop was treated according to common practice and the other part according to MLHD. Herbicide use, level of weed control and yield were assessed for both systems. For example, this kind of testing was done in sixteen ware potato crops in 1998 - 2001. Common practice in the potato crops was an LDS of Sencor (metribuzin) sometimes combined with Basagran (bentazon) or Titus (rimsulfuron). In a few cases a soil

Table 2. *Results from on farm research in 16 ware potato crops in 1998 - 2001 in the Netherlands.*

	Common practice	MLHD
Use of contact herbicides (kg/ha)	0.3	0.2 *
Herbicide efficacy on scale 1-10	7.9	7.7
Relative yield (%)	100	103

* the difference from common practice would have been larger if soil herbicides were also taken into account

herbicide was applied as well. On average 30 % less herbicide was used with MLHD while the level of weed control remained good. Higher yields were observed in MLHD-treated crops. This is explained by a milder effect on the crop of the lower MLHD dose.

The response of weeds and crops to herbicide treatments is species specific. In many cases, we demonstrated that less-sensitive species are less affected by the herbicides, and this can be visualised by the MLHD measurements. An example is available from measurements in an onion crop. Weeds such as *Stellaria media* that are relatively sensitive to Basagran and Actril 200 (ioxynil) yielded a PS1-reading of > 80 2 days after treatment while a species such as *Senecio vulgaris* gave a PS1-reading in the order of 50. The onion crop gave a PS1-reading in the order of 25.

Farmers experiences with MLHD

The on-farm research provided feed back that was used further to develop the MLHD concept. A summary of the feedback is given in Table 3, which is the outcome of an evaluation study in 2002. The evaluation showed that farmers, advisors, and contractors are most interested in the efficacy measurements of MLHD, especially in uncertain situations when they feel that the dose may not have been high enough or when it suddenly rains after a spray. Some farmers also like to measure the response of the crop to the herbicide treatment. The aspects evaluated are ranked in increasing order of usefulness/interest. At present slightly more than 100 MLHD-meters are used in practice in The Netherlands

Table 3. View on aspects of the MLHD concept by users on a scale of 1 (not useful) to 9 (very useful).

Aspect of MLHD	Rating
A system that reduces environmental side effects	4
A system that shows social responsibility	4
A system that minimises costs of herbicides	4
A system that advises on low doses of herbicides	5
A system that predicts herbicide efficacy shortly after treatment	7
A system that visualises the crop response on herbicide use	7

Potato haulm killing and MLHD

Currently there are experiments to test whether the MLHD concept can also be applied for rational use of products for haulm destruction in potato (leaf dessicants). An additional sensing technique is used in the experiments. A Crop-scan meter is used to determine the amount and the activity of the biomass of the potato haulm (canopy) shortly before application. The reading of the Crop-scan meter is used to generate a minimum effective dose of the potato dessicant. The Crop-scan measures reflection of the canopy, which is indicates the biomass (LAI) and its activity.

In addition to Crop-scan, step 2 of the MLHD concept can be applied to predict whether the applied dose will indeed be effective or not. DSR (decision support rules) are available for the registered products in The Netherlands. These DSR however, are not yet in MLHD online, but probably will be in the near future.

MLHD online

The 2002 version of MLHD can be viewed in Dutch on www.mlhd.nl. MLHD online was developed in a joint project with Opticrop b.v. in Vijfhuizen, the Netherlands and the intellectual property of MLHD DSR belongs to Plant Research International b.v.. Opticrop is a company that is specialised in Agro ICT, focused on the Dutch agribusiness. MLHD online version 2002 is still a prototype version.

The core of MLHD online is a spreadsheet model in which the user enters information on the weeds he wants to control. In return, MLHD online advises on dose. A layer of screens guides the user through the model (Figure 4). Firstly, when MLHD online is accessed, a screen with general information on MLHD online is shown. The user can choose between several crops. When potato is chosen, a screen with herbicide options in potatoes and sensitivity of weeds to these herbicides is shown. Upon selection of a herbicide-weed combination, a table with dose-advice based on weed stage is

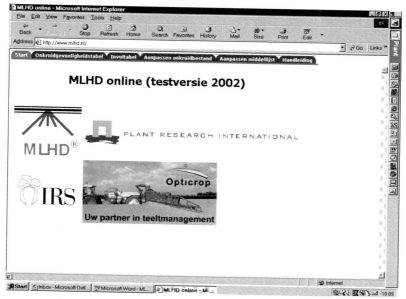

Figure 4a. Screen shot of MLHD online (www.mlhd.nl): opening screen.

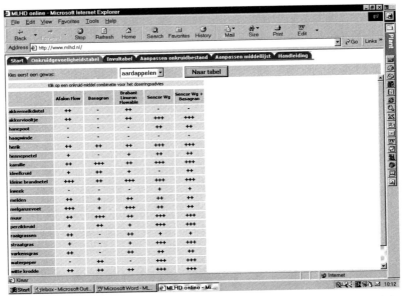

Figure 4b. Screen shot of MLHD online (www.mlhd.nl): weeds - herbicides sensitivity table.

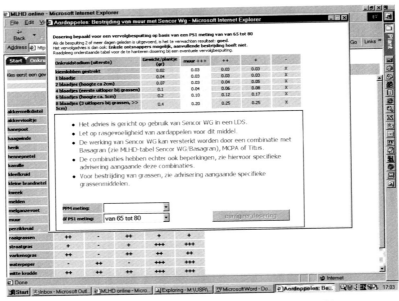

Figure 4c. Screen shot of MLHD online (www.mlhd.nl): dose advice table.

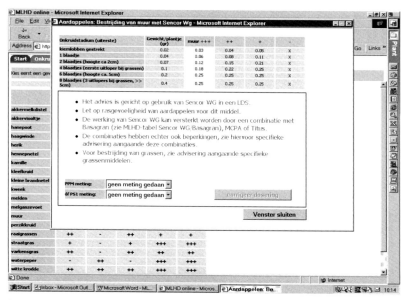

Figure 4d. Screen shot of MLHD online (www.mlhd.nl): dose advice table corrected for PS1-measurements.

presented. If data from measurements of efficacy are available and entered, modified advice will be presented. Some user-friendly options are added to allow rapid comparison of different strategies or situations. Also, text boxes with background information on crops, strategies, weather effects and herbicides are included.

Prospects

MLHD appeals the most to farmers who are interested in either minimising herbicide use while maintaining a good level of weed control, or minimising the effect of herbicides on the crop. Other uses of MLHD are when coupled with the sensing techniques in situations where there is much uncertainty on the efficacy (e.g. rain after application) or in experimental situations (e.g. testing additives). A limitation is that MLHD is only ready for use in practice for photosynthesis-inhibiting herbicides. It is important that the MLHD database is extended beyond these herbicides. R&D at Plant Research International is focused on the extension. For example, DSR are available for glyfosate and glufosinate ammonium applied to annual weeds,. They will be added to MLHD online shortly. R&D on sulfonyl urea compounds is current.

MLHD online will probably be linked in 2003 to a model that takes into account weather effects on herbicide efficacy. At present, such a model, named GEWIS, is available in The Netherlands, from Opticrop b.v (*e.g.* Hoek, 2002). GEWIS advises on the optimal timing of pesticides in relation to 20 weather and product factors.

MLHD requires an investment of time and money by the user. In return, the user saves on herbicide costs and may have a higher crop yield. The investment of time and money is limiting use of MLHD in practise. Farmers that have the time and are interested in optimal crop production are most interested in MLHD. A small group of farmers gets an additional benefit for using MLHD. They produce for an EKO-label (Milieukeur). When they apply certain herbicides with MLHD, they get fewer penalty points. Here MLHD is seen as good agricultural practice. Broadening this perception is an aim of the people involved in the development of MLHD. Most recently, the largest pesticide distributor in The Netherlands, Agrifirm b.v. in Bleiswijk, became involved.

References

Genty, B. & J. Harbinson, 1996. Regulation of light utilization for photosynthetic electron transport. P. 67-99 in: Photosynthesis and the environment. Kluwer Academic Press, The Netherlands. Neil. R. Baker (ed).

Hoek, J., 2002. Software programma's onkruidbestrijding. PPO projectrapport 1236327. Praktijkonderzoek Plant & Omgeving, sector AGV, Lelystad, The Netherlands.

Kempenaar, C., 2002. MLHD handleiding, versie mei 2002 voor internet (in Dutch). www.mlhd.nl.

Kempenaar, C. & H.A.G.M. van den Boogaard, 2002. New insights and developments in the MLHD-concept of weed control. Pagina's 55-59. In: Plant spectrofluorometry: applications and basic research, O. van Kooten & J.F.H. Snel (eds.). Wageningen University, Wageningen, The Netherlands.

Kempenaar, C., R.M.W. Groeneveld, A.J.M. Uffing, R.Y. vander Weide & J.D.A. Wevers, 2002. New insights and developments in the MLHD-concept of weed control. Proceedings of the 12th EWRS symposium 2002, Papendal, The Netherlands, p. 98-99.

Ketel, D.H. & L.A.P. Lotz, 1997. A new method for application of minimum-lethal herbicide dose rates. Proceedings of the 10th EWRS Symposium 1997, Poznan, Poland, p. 150.

Ketel, D.H. & L.A.P. Lotz, 1998. Influence of allocation and detoxification of metribuzin in *Chenopodium album* on the reliability of prediction of the minimum lethal herbicide dose rate. Weed Research 38: 267-274.

Name of the DSS

Ideotyping-potato

Who owns the DSS, Principal author's, address,...

Anton Haverkort, Plant Research International, Wageningen University and Research Centre, P.O. Box 16, 6700 AA Wageningen, The Netherlands

anton.haverkort@wur.nl

What questions does it address?

- what yields are expected when management changes?
- what yields are expected in dry or wet years, with and without irrigation?
- what yields are expected when crop characteristics are changed (e.g. a changed base temperature for development, deeper rooting, stopping leaf growth and death when full ground cover is reached
- which crop characteristics have to be combined to obtain a benefit (e.g. frost resistance and earlier planting are only beneficial when irrigation takes place, if not the soil is too dry during tuber bulking)

What input is needed?

Long runs (30 years) of weather data such as maximum and minimum daily temperatures, precipitation, solar radiation and evapotranspiration; soil characteristics from which rooting depth and water holding capacity can be derived. Long term actual yields of potato to verify model performance.

Who is it for, who are the intended users?

- conventional potato breeders to predict the possible impact of changed crop characteristics
- genetic engineering companies to calculate the impact of far fledged (combinations of) changed genetic properties
- research institutions and commercial companies developing strategies to optimize yields through proper use of genetics, environment and crop management (GxExM). IDEOTYPING POTATO allows a risk analysis before introducing new genotypes into new environments

What are the advantages over conventional methods (whatever those may be)

The conventional method of the identification of ideal genotypes for particular environment consists of screening many genotypes and detecting the best adapted ones. With IDEOTYPING POTATO the optimal set of 'genes' can be calculated before starting a breeding program for such an environment. New areas can quickly be scanned for potential suitability and many years can be analysed for risk of climatic hazards.

How frequently should the DSS be used or its values be updated?

IDEOTYPING POTATO can be used at any time anywhere. Calculating yields can be done when the need for the DSS arises, as long as suitable input data are available

What are the major limitations, and what scientific / technical enhancements would be desirable?

Extending it to quality characteristics requires models and data

13. IDEOTYPING-POTATO a modelling approach to genotype performance

A.J. Haverkort and C. Grashoff

IDEOTYPING-POTATO is a decision support system for potato breeding. An ideotype is the ideal genotype for a particular environment. A potato growing season is characterised by a temperature window (neither too cold nor too hot), and resources such as solar radiation, water and nutrients. The ideotype has a growth cycle with a length that matches the length of the available growing season. The crop growth model LINTUL contains many parameters that represent genetic properties of varieties that allow such matching. Adapting parameter values to optimise yields mimics the alteration of a genetic trait. Examples are the sprout growth rate and the leaf area expansion rate. Also factors with a greater impact that may be achieved through genetic modification can be explored. What happens if the crop stops making new leaves once full ground is cover achieved?. Changing the values of model parameters is similar to subjecting a genotype with a changed property to a target environment. With the decision support system IDEOTYPING-POTATO the optimal set of 'genes' can be calculated before starting a breeding program for any environment. New areas can quickly be scanned for potential suitability and data from many years can be analysed for risk climatic hazards.

Introduction

Ideotyping is the action of looking for the **ide**al gen**otype** for a particular environment. The suitable environment for potato production is determined by the suitable temperatures for growth. Breeding more adapted and higher yielding varieties - besides management - is considered the best option to increase yields in an area. The creation of improved varieties may be through conventional breeding or through genetic modification (GM). Before starting such a breeding programme, a breeder could usefully conduct an ideotyping exercise to answer the following kinds of questions so as to give guidance on the alterations to characters that might achieve the aim of the breeding programme:

- If a crop is planted substantially earlier how many degrees frost resistance is required then?. In other words if a genotype has a certain extra degree of frost resistance how much earlier can the crop be planted without a risk of freezing?
- The sprout growth rate at planting, planting depth and soil temperature determine the emergence rate after planting. If a genotype were created with a substantially

greater sprout growth rate how much earlier would such a genotype emerge and what would the repercussions be for yield?

- Grasses start growing in spring at lower temperatures than potato. Maize needs a still higher base temperature for growth. How much faster would a potato crop develop if its base temperature for growth were lowered?

- When a potato emerges it develops its foliage through two processes. One is temperature driven whereby dry matter is translocated from the mother tuber to the sprout and emerging crop. Once the foliage develops it intercepts radiation and produces dry matter, part of which benefits the foliage. The temperature driven foliar development depends on the initial leaf area at emergence and on the subsequent relative leaf area development rate. What happens to the leaf area development (and light interception and dry matter production) if genetically altered crops have higher initial leaf areas (or lower) at emergence and / or higher leaf area development rates?

- The potato originated in the tropical highlands with high rates of incident solar radiation. In temperate climates solar radiation is less intense requiring thinner leaves. If leaves were substantially thinner to what degree would that positively influence yields?

- Potato differs from cereals in many aspects. One of them being that the tubers are stored under ground. This makes it unnecessary to produce straw holding the grains in the air. Another major difference is that potato drops its lower leaves once a leaf area index over about four is reached whereas cereals keep their flag leaf and no leaf shedding takes place. When nitrogen and carbon is needed for the storage organs, lower potato leaves translocate it and are shed while maintaining near optimal nitrogen levels in the upper leaves whereas cereal leaves contain less and less nitrogen towards crop maturity. An ideal genotype of potato would produce only about 4 leaf layers and then not drop leaves and residual carbon and nitrogen with them. What happens to potato crops when they only achieve a maximum LAI of say 4 and than either senesce like cereals, i.e. empty the leaves of nitrogen, or not.

- Potato crops die because eventually during crop development the tubers absorb all daily accumulated dry matter leaving none for the foliage that then dies after some time. The moment of tuber initiation (when the plant has accumulated a certain amount of dry matter) and subsequent dry matter allocation to the tubers of all dry matter produced per day determine the onset and subsequent rate of senescence. Varying the genetic influences will vary the moment and pattern of senescence and the adaptation of the crop to its environment.

- Potato crops tend to be relatively sensitive to drought. Many crops root more deeply and are able to recover sooner after a dry spell. Sugar beet is an example of a crop performing better than potato in response to drought. Potato on the other hand has a higher water use efficiency producing more harvestable dry matter per unit transpired water than cereals, especially rice. Rice is grown in water leading to high evaporation rates and all cereals have low harvest indices due to production of straw containing lignin. How would deeper rooting, and the associated increased soil water availability, influence the performance of a potato genotype? And what would be the influence of an increased 'water use efficiency' per se? It is also of interest to know what would happen if the potato crop had a more cereal-like behaviour: not dropping its leaves (so more transpiration and respiration during drought) but keeping them necessitating recuperation

- The light use efficiency of potato crops tends to be stable and may be influenced only by the light intensity (light saturation may occur) and temperature (very low temperatures reduce photosynthesis and very high temperatures increase respiration. No genetic variability is expected.

Runs made with IDEOTYPING-POTATO are based on current physiological opinion but it is inherent in the flexible selection of characters and parameters that those chosen could be physiologically incompatible and so one cannot be certain that the results will reflect an achievable physiological basis.

The science behind the DSS

Potatoes cannot withstand frost although when frost happens shortly after emergence re-growth is possible. When temperatures are too high potato production is not possible either. So, the length of the growing season of potato crops grown in temperate climates is determined by the ambient temperatures. Emergence takes place when the greatest chance of night frost has passed. The harvest takes place before the risk of night frost in autumn has become too high. Earlier planting and later harvesting would lead to adverse weather conditions even if frost would not damage the crops. Often in early spring the land is not accessible due to excess moisture in the soil, and in autumn harvest may be made impossible due to rainfall. If, however, genotypes were developed that withstand low temperatures, it may be assumed that mechanization techniques would be developed to overcome the early and late soil moisture conditions. In high temperatures the respiration rate increases thus reducing total dry matter accumulation. Moreover, an increased proportion of the dry matter produced is allocated to the foliage, leaving less for the tubers. And, higher temperatures lead to a reduced dry matter concentration in the tubers. When neither yield limiting factors (e.g. water, nutrients) nor yield reducing factors (e.g. pests, diseases) are present, the potential tuber yield is determined by the variety planted, the temperature, the daylength, the amount of solar radiation and the concentration of carbon dioxide in the air. Temperature and daylength determine the development of the crop, i.e. the emergence, the initial leaf development, tuber initiation, dry matter distribution over the various plant parts and leaf death. Besides temperature, the total amount of dry matter accumulated by the crop is determined by solar radiation (photosynthetically active radiation between 400 and 700 nm) and the concentration of carbon dioxide in the air (about 350 ppm). It is a particular feature of the yield-defining factors that once the grower has planted the seed potatoes, he/she can no longer exert any influence on them. The version of the LINTUL model used here to calculate potential yields based on temperature and solar radiation is described by Kooman & Haverkort (1995) and the effect of drought by van Keulen & Stol (1995)

To establish a DSS on ideotyping and to enable it to work beside a proper mechanistic model (LINTUL-POTATO), long term weather and soil data are needed for the target area. In this paper the starch potato production area in the north east of the Netherlands is elaborated as a case study.

Description of the target area

For an ideotyping study a growing season is targeted where the ideotype fits. A description of the growing season has two aspects: spatial (where will the ideotype be located, what are the soil and long term weather conditions) and temporal (what is the fluctuation in weather and what are the risks of e.g. drought and night frosts?). The starch potato area in the North East of the Netherlands, where about 5,000 ha of starch potatoes are grown annually, is used here to illustrate the approach.

The soil types are peaty and sandy. The area has lower yields on average than the clay soil areas in the Netherlands. The main problems are late spring frosts and early autumn frost reducing the length of the growing season. The water holding capacity of the sandy soils is less than that of loamy soils. Finally the rooting depth of the soils is often reduced due to alluvial sandy or acid peat layers at shallow depths. Irrigation is not very common as the rate of return on the investment in such equipment is too low to justify the capital allocation. The yield limiting and reducing factors in the starch potato area in the North East of the Netherlands are the following: late frosts occur frequently in May and early June, the water holding capacity of the soils is limited due to shallow sandy and peaty soils, and, besides late blight, the occurrence of potato cyst nematodes often seriously affects yield. Potatoes are grown in very short rotations of only two years with either sugar beet or wheat. Figure 1 shows the long-term average mean daily temperature, radiation and precipitation deficit (= evapotranspiration precipitation) during the growing season and the average yields of the whole region. Plotting yields against temperature, radiation and precipitation deficit did not yield any significant correlation

The average daily temperature during the growing season (Figure 1A) varies from 14 °C in the wet and cloudy year 1962 to 18.5 °C in the dry and sunny year 1990. Similar patterns are seen in the data for radiation (Figure 1B) and precipitation 'deficit' (Figure 1C). Dry years may have a deficit of over 2.5 mm day^{-1} whereas wet years may have a net surplus throughout the growing season. There is no relation at all between yield and incident radiation (Figure 2A) nor between yield and precipitation deficit. This is probably due to compensation: dry years have less rain but more sunshine.

To determine the effect of changed parameter values of the crop growth model LINTUL-POTATO to mimic genetic variability for this trait it is better to run the model for individual years to capture the influence of the variation in weather than with the long term average values. In the case study with starch potato two years were chosen: a suny and dry year (1976) and a much wetter and overcast year (1984). The model was run with (not irrigated) and without (irrigated) drought stress with the default parameter values as shown in the next section. Figure 3 shows examples of the results of the model

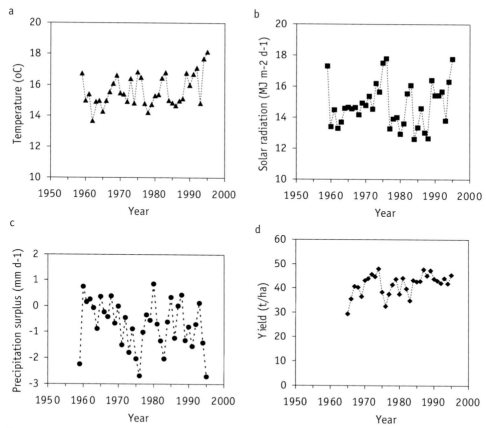

Figure 1. Eelde airport long term temperature (A), solar radiation (B), precipitation surplus and deficit (C) between June 15 and September 24 and average regional starch potato yields (D) in the starch potato region in the North East of the Netherlands.

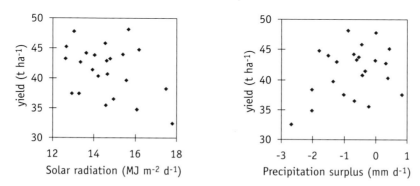

Figure 2. Relations between solar radiation during the growing season and yield (A) and precipitation deficit and yield (B). Starch potato regional data 1965-1988.

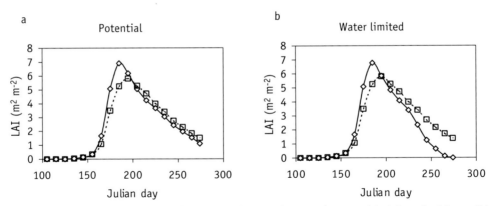

Figure 3. Simulated LAI in a dry (◊1976) and a wet (□ 1984) year with (a) and without (b) irrigation.

calculations of leaf area index (LAI) and Figure 4 of tuber dry matter yield in a screenshot. Obviously the highest yields were simulated with irrigation (4a) especially in the sunny year and the lowest yields without irrigation, especially in the dry year (3b). The weather data of these years were used in the ideotyping model runs for the 'wet' and 'dry' years.

Ideotyping runs

The ideotype of a potato with higher potential and attainable yields than the present ones can be approached hypothetically by earlier planting of seed potatoes with more and longer sprouts. These sprouts should grow at lower soil temperatures, the sprouts emerge with a higher foliar mass and with thinner leaves that have a few degrees of frost resistance. The leaves should expand more rapidly starting with a lower base temperature and the resulting foliage should consist of leaves with such a distribution pattern and position of the leaves that more light is intercepted at any given LAI. If the crop were to suffer less from nematodes and drought through improved resistance and deeper rooting respectively, then the light would be converted into dry matter more efficiently and, also, crop longevity would be increased. If the crop would produce only enough new leaf layers to cover the ground fully **with** green leaves then more of the total dry matter would be partitioned to the tubers.

When using the crop growth model LINTUL for ideotyping, many of these changes can be 'implemented' easily by varying a number of parameters as in the following examples. The bracketed sets of three values show what crop and management parameters were varied in the ideotyping runs with 30-year weather data and the figure in **bold** shows the default value in the model that is assumed to be representative for the local climatic conditions, the local varieties and local management

- the date of planting (DP): this date may vary to show the frost risk the present genotypes are subjected to indicate what happens if the ideotype is more frost resistant (DP = March 26, **April 15,** May 5)
- planting depth (PD): placing seed deeper may protect tubers from light and diseases and place them more favourably close to resources (PD = 8 cm, **12 cm,** 16 cm)
- the (relative) sprout growth rate (SGR): longer sprouts at planting with a faster growth rate will emerge sooner (SGR = 0.5, **1.0,** 2.0 cm dd^{-1})
- base temperature (BT): below 2 °C the crop does not develop. No sprout growth takes place nor leaf expansion, photosynthesis nor respiration. A lower base temperature would speed up the development (BT = 0, **2.0,** 4.0 °C)
- planting density (PlD): presently plants are grown in rows 75 cm apart with a planting distance of 30 cm within the row. Higher plant densities lead to increased seed costs but to earlier canopy closure (PlD = 3, **4,** 6 plants m^{-2})
- number of stems per plant (SPP): with 4 plants per m^{-2} and 4 main stems per plant the desired 16 stems per m^{-2} are achieved. Ideotypes generating more stems per plant may lead to earlier canopy closure and higher yields (SPP = 2, **4,** 8)
- the foliar mass at emergence (FME): thin sprouts will emerge with not much foliage weight. The leaves initially are thick and folded and will expand and grow thinner with thermal time. The more foliar mass at emergence the earlier the canopy will close. At 50 % emergence, leaf weight was found to be 0.08 g stem^{-1} (FME = 0.03, **0.08,** 0.4)
- specific leaf area (SLA): potato plants in the area usually have an SLA of 250 cm^2 g^{-1} leaf dry matter. The canopy will intercept all the incident radiation sooner if leaves are thinner (SLA= 150, **250,** 350 cm^2 g^{-1})
- light extinction by the foliage: the relation between LAI and the proportion of ground covered is about 1:3, i.e. at an LAI of 3 the crop fully covers the ground with green leaves. If the position and orientation of the leaves were such that the soil would be covered at lower LAI values that would improve light interception by the crop (f$_{int}$) through an increased light extinction coefficient (k) as follows from Beer's law: f$_{int}$ = 1- e^{-k*LAI} (k = 0.7, **1.0,** 1.3)
- relative leaf expansion rate (RLER): when the foliage initially develops after emergence through transfer of dry matter from the mother tuber, this is a temperature driven process up to an LAI value of about 0.75 with a relative leaf area expansion rate of 0.10 m^2 added to each existing m^2 of leaf area per degree-day. After this, dry matter allocation to the foliage is the driving force for growth of leaf area (RLER = 0.006, **0.010,** 0.016 m^2 m^{-2} dd^{-1})
- onset of leaf death: the average maximum age of a leaf that has sufficient water and minerals available is about 945 degree-days. This is a value valid for the late starch potato varieties that predominate in the target area. Lack of water and minerals, and shading by higher leaves cause earlier senescence. Also, earlier maturing varieties have shorter living leaves as the tubers compete for available assimilates. LINTUL also reduces leaf longevity with increased LAI due to shading effects that cause plants to reallocate dry matter to upper leaves (leaf longevity =725, **945,** 1100 dd)

- allowing a maximum LAI only: if a maximum value is imposed on the LAI the amount of dry matter not invested in the canopy is allocated to the tubers. If the model does not allow leaf senescence there is an additional benefit because no new leaf layers need to be formed (maximum LAI = 3, **default (free),** 6)
- light use efficiency (LUE): the standard LUE = 2.7 g dry matter produced per MJ photosynthetically active radiation intercepted by the canopy that covers the ground. The model in the ideotyping runs considers 10 % higher and lower LUE values (LUE = 2.4, **2.7,** and 3.0 g MJ^{-1})
- moment of tuber initiation (TI): at 415 degree-days after emergence tubers are initiated and after another 430 degree-days all assimilates produced per day are allocated to the tubers and the foliage then dies. Earlier tuber formation makes the genotype earlier. Where in the model runs TI is earlier or later, the onset of leaf senescence is also brought forward (or backward) by the same number of degree-days (TI = 15, **415,** 600 dd after emergence)
- the (relative) tuber growth rate (TGR): after tubers are initiated dry matter is allocated to them from the daily growth. The absolute tuber growth rate depends on daily growth and proportion allocated. Depending on the proportion of dry matter partitioned to the tubers earlier or later in the growing season the moment will arrive when all the dry matter produced is allocated to the tubers. That is when leaf senescence starts. In simulation runs that moment (resulting from TGR) is imposed to identify ideotype values (TGR leads to 100 % dry matter allocation within 185, **430,** 685 dd)
- rooting depth (RD): the model assumes an average maximum rooting depth of around 50 cm. Deeper rooting would mean a greater access to water during the growing season, therefore a run with rooting depth of 80 cm is included (RD = 30, **50,** 80 cm)
- influence of drought on rate of leaf death (LDR at permanent wilting point, PWP): the ratio of actual : potential evapotranspiration (Ta/Tp) is a measure of the drought stress the crop is subjected to. In the standard runs the leaves die at a rate of 0.05 m^2 m^{-2} dd^{-1} when Ta/Tp = 0 (PWP). When the soil water content is nearing field capacity there is no leaf death caused by drought *per se* (LDR = 0.02, **0.05**, 0.08 m^2 m^{-2} dd^{-1})
- water use efficiency (WUE): WUE is expressed as dry matter produced per litre of water transpired. Potato is assumed to be able to assimilate during part of the day with partially closed stomata. The model in the ideotyping studies assumes a varying relative crop transpiration rate (RTR) as a proportion of the daily open pan transpiration. (RTR = 0.6, **0.9**, 1.3)

Results of model runs

Table 1 shows the simulated production of tuber dry matter, harvest index and intercepted photosynthetically active radiation of crops grown in dry and wet years with and without irrigation. The table shows the results of the model run with default values

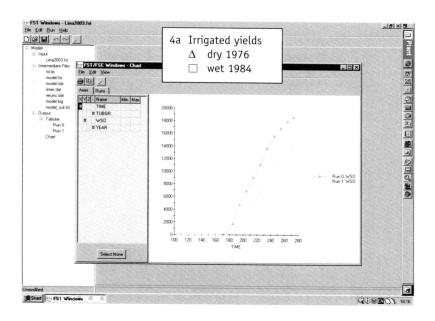

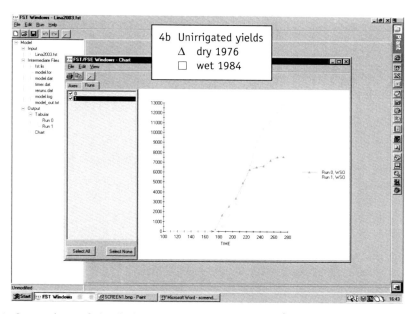

Figure 4. Screenshots of simulation runs of tuber yield (g m⁻² dry matter).

and with examples of three adapted parameter values: adapted base temperature for development, adapted LAI-development (leaf growth stopped at LAI=3 and 6 and no leaf death took place) and adapted rooting depth. The standard default setting (see also

Table 1. Simulated tuber dry matter production, harvest index (HI) and intercepted photosynthetically active radiation (PAR) of crops grown in dry and wet years without (-) and with (+) irrigation. Examples of three adapted parameter values.

Parameter values	Type of season		Tuber DM production	HI	Intercepted PAR (MJ m^{-2})
default	Dry	-	7.4	0.59	883
default	Wet	-	12.4	0.74	668
default	Dry	+	19.0	0.77	954
default	Wet	+	12.8	0.75	670
Tbase =0 °C	Dry	-	8.1	0.62	875
	Wet	-	13.2	0.78	663
	Dry	+	20.1	0.80	970
	Wet	+	13.6	0.79	669
Tbase = 4 °C	Dry	-	7.1	0.58	868
	Wet	-	10.2	0.68	600
	Dry	+	16.6	0.73	898
	Wet	+	10.5	0.69	601
LAImax =- 3	Dry	-	11.1	0.78	954
(leaves stay alive)	Wet	-	14.0	0.84	663
	Dry	+	21.2	0.87	948
	Wet	+	14.3	0.84	663
LAImax = 6	Dry	-	9.7	0.69	986
(leaves stay alive)	Wet	-	13.3	0.77	690
	Dry	+	20.9	0.82	986
	Wet	+	13.7	0.78	690
Root depth 30 cm	Dry	-	5.3	0.56	848
	Wet	-	9.6	0.69	650
	Dry	+	19.0	0.77	954
	Wet	+	12.8	0.75	670
Root depth 30 cm	Dry	-	12.1	0.69	907
	Wet	-	12.8	0.75	670
	Dry	+	19.0	0.77	954
	Wet	+	12.8	0.75	670

Figure 3) calculated the lowest tuber DM production in the dry unirrigated condition and the highest tuber DM production in the same year but irrigated. The two other conditions showed intermediate results. The low harvest index following the sunny dry season shows that initially the crop suffered mainly from drought in the second part of the season. Lowering the base temperature increased tuber DM production substantially in all the four seasons (>6%). Restricting the LAI to 3 only increased the yield in the sunny dry season by about 50 % largely due to the increase in harvest index and less due to the increase in intercepted radiation. Restricting the LAI here to 6, increased tuber DM production by about 30 %. All other situations led to an increase of about 10 %. An increase of the rooting depth from the standard 50 cm to 80 cm increased the yield in the dry sunny year by more than 64 %. In the wet, overcast year and in the irrigated conditions deeper rooting had no effect on yields.

Table 2. Response of tuber DM production when favourably adjusting parameter values of a standard crop grown in dry and wet years with and without irrigation (- = negative response, 0 = no response + = > 2% increase in tuber DM production, = ++ = > 5% increase in tuber DM production).

Altered crop characteristic	Unirrigated		Irrigated	
Type of growing season	Dry	Wet	Dry	Wet
Planting date	0	+	+	+
Planting depth	0	+	+	+
Sprout growth rate	0	++	+	++
Base temperature for development	++	+	+	+
Plant density	0	0	0	0
Number of stems per plant	0	0	0	0
Leaf weight at emergence	0	0	0	0
Specific leaf area	++	0	0	0
Light extinction coefficient	0	0	0	0
Leaf expansion rate	0	0	0	0
Onset of leaf death	++	0	0	0
Maximum LAI, leaves die	++	-	—	-
Maximum LAI, leaves stay alive	++	++	++	++
Light use efficiency	++	++	++	++
Moment of tuber initiation	-	-	-	-
Tuber growth rate	++	++	++	++
Rooting depth	++	+	0	0
Drought dependent leaf death	++	0	0	0
Water use efficiency	++	+	0	0

All the other simulation runs with adapted parameter values mimicking genetic traits are shown in Table 2. Here the tuber yield response is given when favourably adjusting parameter values of a standard crop grown in dry and wet years with and without irrigation. A number of 'improvements' did not influence tuber DM production under any condition. Increased planting density, number of stems per plant, sprout leaf mass at emergence, light extinction coefficient and leaf expansion rate did not lead to increased tuber DM production. Apparently, most of the factors leading to earlier canopy closure only very marginally contribute to this aim with only a few days earlier closure, not enough to be reflected in radiation interception and tuber DM production. Changing the moment of tuber initiation (without varying other crop characteristics) invariably reduced tuber DM production whereas increased tuber growth rate invariably increased tuber DM production. Some factors only increased DM production in the dry sunny season such as increased specific leaf area, earlier leaf death (and the restricted LAI at 3), and, of course, the parameter values that influence the water relations such as the water use efficiency and rooting depth. The factors leading to an earlier emergence of the crop such as earlier and shallower planting and faster growing sprouts only increase yields when, later during the growing season, water does not limit growth.

IDEOTYPING-POTATO allows a quick exploration of the physiological and ecological adaptation of the potato crop. The adaptation of a crop characteristic usually only partly exerts an influence on final tuber DM production because the factor does not necessarily influence the development of yield directly. This may be because an early action is diluted e.g. thinner leaves lead to earlier canopy expansion but light extinction early on is reduced and overlap of leaf layers starts earlier. Early crop development through earlier emergence will lead to an earlier exhaustion of the water supply in summer and will only be beneficial when accompanied by deeper rooting and/or irrigation. Changing only one gene therefore seems hardly an option and usually a set of genes need to be altered. The results of model runs of several scenarios are shown in Table 3. Compared to the default setting the leaf area development is hastened by increasing the initial leaf area per stem from 0.002 to 0.01 m^2, by increasing the relative leaf expansion rate from 0.01 to 0.016 m^2 m^{-2} dd^{-1} and by increasing the specific leaf area from 250 to 350 cm^{-2} g^{-1}. The crop emerges 10 days earlier through earlier and shallower planting and an increased sprout growth rate. The crop resists 2 degrees lower temperature than the default crop. Its LAI does not exceed the value of 4 and leaves do not die but remain green until final harvest. Unirrigated crops grown in dry and sunny years especially benefit from the proposed changes (+ 78%) but also in other years and with irrigation the expected increases in tuber DM production exceed 40 %.

Future developments

Improvements in IDEOTYPING-POTATO will be a more detailed mechanistic description of crop growth and development. This will increase the number of genes that are mimicked. Examples are parameter values related to photosynthesis and respiration, fine-tuning of dry matter distribution over a greater number of organs, distinguishing

Table 3. Simulated tuber dry matter production, harvest index (HI) and intercepted photosynthetically active radiation (PAR) of crops grown in dry and wet years without (-) and with (+) irrigation. Defaults settings and optimisation of 9 parameter values.

Parameter values	Type of season		Tuber DM production	HI	Intercepted PAR (MJ m^{-2})
default	Dry	-	7.4	0.59	883
default	Wet	-	12.4	0.74	668
default	Dry	+	19.0	0.77	954
default	Wet	+	12.8	0.75	670
Optimal setting	Dry	-	13.2	0.78	1228
	Wet	-	17.8	0.83	875
	Dry	+	26.9	0.87	1229
	Wet	+	18.3	0.84	875

between carbon and nitrogen and the spatial arrangements of organs, especially leaves and roots. The influence of biotic and abiotic factors can be included not only to optimise the yields under optimal conditions but also under sub-optimal conditions. This will also allow the exploration of the resource use efficiency (water, nutrients and biocides). Finally, model development is needed to cover quality aspects such as tuber size distribution and dry matter concentration. Ideotyping in areas where no potatoes are grown will require a follow-up with genotypes approaching the ideotypes to assess the accuracy of the model.

Acknowledgements

The case study of IDEOTYPING-POTATO on starch potato was funded by AVEBE. The contribution of this potato starch company is gratefully acknowledged.

References

Kooman, P.L. & A.J. Haverkort, 1995. Modelling development and growth of the potato crop influenced by temperature and daylength: LINTUL-POTATO. In A.J. Haverkort & D.K.L. MacKerron (Eds.) Potato Ecology and modelling of crops under conditions limiting growth. Kluwer Academic Publishers pp. 41-59.

Van Keulen, H. & W. Stol, 1995. Agro-ecological zonation for potato production. In A.J. Haverkort & D.K.L. MacKerron (Eds.) Potato Ecology and modelling of crops under conditions limiting growth. Kluwer Academic Publishers pp. 357-371.

14. Necessity and sufficiency or the balance between accuracy and practicality

D.K.L. MacKerron

"A theory has only the alternatives of being right or wrong. A model has a third possibility. It may be right but irrelevant."
                                                                                    Manfred Eigen

In developing computerised Decision Support Systems (DSS), the science behind the component models must be right and, just as importantly, the science must be appropriate. The features and functions should be enough for the task in hand, but no more than enough - necessary and sufficient. What is enough depends a lot on what the DSS is intended to do and, also, on the level of accuracy required. These have consequences for the level of practicality and so of acceptability of the DSS. Scientists have a tendency to try to include all that they know about a problem but this can lead to 'unnecessary' accuracy from the users' viewpoint. Another tendency of modellers is to 'believe' their own models, forgetting the inbuilt errors of the component algorithms. This chapter begins by reviewing the unavoidability of errors to re-awaken awareness among modellers and DSS developers that each answer given by our semi-mechanistic models is only one among a population of right answers. It then continues with a consideration of features of DSS on potato where the decision on what is necessary and what is sufficient needs careful consideration.

Introduction

Decision support systems (DSS) are usually based on one or more models that describe a process or a state, the 'target'. These models are generally abstractions of reality, simplifications, and as such they do not and cannot encapsulate all factors and all influences on the target. The best that a modeller can aspire to is to include enough of the important factors and variables to allow his or her model to describe the target adequately. And this leads to the key question of what is 'adequate'. The Influence Diagram prepared during the development of MAPP (Chapter 8, Figure 3) makes the distinction between influences that are both recognised and represented within the DSS and those influences that are 'recognised' but are not handled - ignored. The possible reasons for ignoring a recognised influence will be discussed.

There is a tendency for scientists to make their models 'better and better'. That is to say, to make them include more and more of the scientist's knowledge with the intention that the models should describe their target with ever increasing precision. However, such precision may be illusory and, particularly for decision support, it may be irrelevant. An enhancement to a model may be considered illusory if it refines the definition of one part of the system or a state variable, such as number of tubers, in a way that is either additive or multiplicative with another part that has a coarser precision and that is not modified by the enhancement. Similarly, where the enhancement entails acquiring values for extra variables and / or parameters, it may be thought ill advised if the cost of the extra measurements is not outweighed by the value of the increased precision.

Over the last twenty years, several computerised models and DSS have been built for a range of crops. Most have failed. - But not in computation. - The farming community simply did not adopt them. Why? There are two outstanding criteria for success or failure. - A DSS must do the job well and it must be easy to use. Most of the earlier attempts at DSS did not focus on the user. They did not recognise that the user is almost certainly not a modeller nor a computer enthusiast. Nor did they recognise that the user's need to provide inputs to the DSS incurs costs in both time and money so, the inputs must be as few as possible and easy to obtain. And the DSS should have a view to the true value of the outputs to the user, that is, to the difference to the user between using the DSS and not using it.

This chapter considers the questions of necessary and sufficient variables and the balance between accuracy and practicality. Where possible, it draws on examples taken from the models and the DSS presented in this book.

The problem of accuracy, and uncertainty
(Modellers aim for precision and forget errors)

All models are falsifiable. That is, it is possible, in principle, to prove them wrong. And so it should be, for models deal with only a small part of reality and even the best models have shortcomings. At best they provide us with a sketch-map of what happens in reality. The building and use of models bring some dangers with them. Among these is the belief of the modeller that his model is right.

A guiding principle in modelling to develop DSS is that the inputs to the models should be as simple and as generally available as possible. This has led to the 'top-down' approach. That is, the models draw on empirical relations derived at as high a level of organization as can be made to work. Lowering the level of organization from which relations are drawn is often said to make a model more 'mechanistic' but it can also be seen as simply pushing back the point of empiricism. Frequently these relations are quantified using the statistical technique of regression analysis.

Regression analysis is usefully applied in a broad set of circumstances when a random variable, Y, called the *dependent variable,* has a mean that is a function of one or more random variables x_1, x_2, ..., x_k, called *independent variables.*

Many different mathematical functions can be used to model a response that is a function of one or more independent variables. These can be classified into two categories: deterministic and probabilistic models. For example, suppose that x and y are related according to the equation $y = \beta_0 + \beta_1 x$ (where β_0, β_1 are unknown parameters). This is a deterministic mathematical model because it does not allow for any error in predicting y as a function of x. We might imply that y always takes the value $(\beta_0 + \beta_1 5.5)$ whenever x = 5.5.

Suppose that we collect a sample of n values of y corresponding to n different settings of the independent variable x and that a plot of the data looks like that shown in Figure 1. It is quite clear from the figure that the expected value of Y may increase as a linear function of x but that a deterministic model is far from a complete description of reality. Repeated experiments when x = 5.5 would give values of Y that vary in a random manner. This tells us that the deterministic model is not an exact representation of the relation between the two variables. Further, if the model were to be used to predict Y, when x = 5.5 the prediction would be subject to some unknown error. - Yet this very point is generally overlooked in what are known as 'semi-mechanistic' models. - This leads to the use of statistical methods. Predicting Y for a given value of x is an inferential process and we need to know the properties of the error of prediction if the prediction is to be of value in real life.

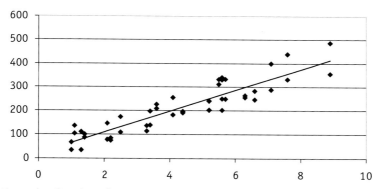

Figure 1. Example of a plot of x versus y.

In contrast to the deterministic model, statisticians use probabilistic models. For example we might represent the responses in the diagram, not by the equation $y = \beta_0 + \beta_1 x$ but by the equation:

$y = \beta_0 + \beta_1 x + \varepsilon$

where ε is a random variable with a specified probability distribution with a mean of zero. Now think of Y as the sum of a deterministic component E(Y) and a random component ε. This model accounts for the random behaviour of Y exhibited in the figure and provides a more accurate description of reality than the deterministic model. Further, the properties of the error of prediction for Y can be derived for many probabilistic models.

Figure 2 represents the probabilistic model $y = \beta_0 + \beta_1 x + \varepsilon$. When x = 5.5 there is a population of possible values of Y. The distribution of this population is indicated on the main portion of the graph and is centred on the line $E(Y) = \beta_0 + \beta_1 x$ at the point x = 5.5. This population has a distribution with mean $\beta_0 + \beta_1(5.5)$ and variance σ^2.

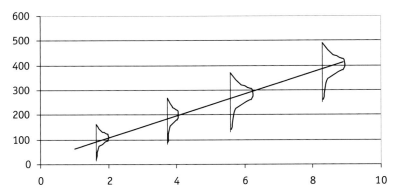

Figure 2. Probabilistic model of the plot of Figure 1.

When x = 7, there is another population of possible values for Y. The distribution of that population has the same form as the other population, the same variance σ^2 but with a mean value of $\beta_0 + \beta_1(7)$. The equivalents are true for each possible value of the independent variable, x.

It is quite normal or usual for scientific textbooks and publications to use deterministic models of reality. Many of the functions that appear in textbooks of physics, e.g. Newton's law relating the force on a body to its mass and its acceleration (P = mf), are deterministic mathematical models of nature which, for practical purposes, predict with little error. In contrast, many models, if not most models, published in biological and other scientific journals are often poor ones by those standards and the scatters of points around the fitted lines, similar to the first figure earlier, show that they are fairly inadequate. Unfortunately, the tendency is not to emphasise the uncertainty underlying the relations that are described. It is, even, to ignore that uncertainty. Yet the distribution of the points for each value of x is as important a part of the relation between Y and x as is the deterministic equation. None of the DSS presented in this book attempt to represent the uncertainty in biological and agricultural observations, although some do at least try to accommodate it by including field observations to re-scale the modelled results.

These points are made simply to stress that when we arrive at a deterministic model using regression analysis we really ought to appreciate and remember the variance in the underlying data and, therefore, the error in any prediction that we make. (Here, of course, 'error' does not mean 'mistake' but difference from what was predicted). Most DSS are based on such relations, and the users will see these errors or discrepancies. A real danger is that they will prevent the user developing confidence in the DSS. Also, the scale of those errors should guide us in deciding whether or not particular 'refinements' to a model are worth making. If the coefficient of variation (CoV) of an observed variable, Y, were, say, 15% around the deterministically simulated value then a development that would alter Y by 5% without modifying the CoV would be unrewarding.

What is necessary and what is sufficient?
Cases drawn from some of the DSS described in this book

Cultivars can cause confusion

In commercial practice the differences between cultivars are an extremely important issue and most countries produce a national list of potato cultivars (e.g. National Institute for Agricultural Botany, NIAB) that generally gives a rating for a number of the characteristics of a cultivar. These include a scaled value for resistance to late blight. So, a DSS such as **PlantPlus** (Chapter 11 - Raatjes et al.) can use such a scale to qualify the performance of a crop of a known cultivar. Differences between cultivars offer a different problem in the model of **PCN development** (Chapter 9 *(*Elliott et al.) as only a few cultivars exhibit even partial resistance to *Globodera pallida*.

Apart from the scores for resistance to pests and diseases, few of the national list ratings quantify any of the features that are parameterised in models of growth and development or of crop nutrition.

The approach of Booij et al. (Chapter 3) does not include differences between cultivars. That of Chambenoit et al. (Chapter 4) used data from many cultivars in deriving and testing relations but appears to treat them all as one. Goffart et al. (Chapter 5) use a cultivar coefficient that has been properly evaluated for the one cultivar, Bintje, and are now facing the problem of characterising others. They suggest values for five other cultivars. Jamieson et al. (Chapter 6) allow several growth parameters to be altered between cultivars. These include durations of growth phases, rate of canopy expansion, maximum green area index, and water extraction (rooting). Yet there is no mechanism proposed, other than extensive trialling, for providing cultivar-specific values for those parameters.

Unavailability of growth parameters on all but a few cultivars means that either the developer of a DSS must settle for a 'standard' set that will describe performance of a typical cultivar. Or else a default set must be provided , with the option to change any of the parameters that the user may know. That is a recipe for confusion and a hindrance to the user unless he happens to grow the cultivar used to provide the default values. It almost certainly will destroy the normal user's confidence in the system as it is saying to the user, "These parameters are important. You should change them for each cultivar". The first option, the standard set, is the better and will allow the user to observe any systematic discrepancies.

In **MAPP** (Chapter 8 - MacKerron et al.) differences between cultivars are an important issue. In the parts of MAPP that handle seed rates and tuber size distribution MAPP uses eleven parameters to cover four areas of development: the size-to-weight relation of tubers, linking grade in millimetres (square riddle) to size in grammes; yield versus stem density; number of daughter tubers as functions of plant spacing and seed size; and the relative variability of tuber size. It is not possible to derive the values of these parameters for all the cultivars experimentally. Instead, a few parameters have been taken to be constant across cultivars and the experimentally derived values of the others were regressed on NIAB ratings for characters such as "shape" and "tuber number". This allowed individual sets of parameter values for over one hundred cultivars to be included in the DSS. Further cultivars can be added to the database within MAPP by likening the shape and the multiplication rate of the new cultivar to those of known ones.

That approach is approximate but it is practical. It is easy for the user to 'add' a new cultivar to the list and it produces results that differ between cultivars in a manner that the user can understand. - Cultivar A has a shape like that of Cultivar B so it has a similar relation between tuber weight and size in millimetres.

Photoperiod?

The length of the day during growth (photoperiod) has effects on the potato crop. - Principally these concern timing of tuber initiation and number of tubers formed, although the differences in timing of tuber initiation can have knock-on effects on yield and, so, on tuber size distribution, by changing the time available for tuber growth.

Temperature and photoperiod interact in their effects on tuber initiation. When the potato was first introduced to Europe, tuberisation was strongly influenced by photoperiod, initiation requiring short days. Modern cultivars are much more day neutral and so over much of Europe the effect of day length on potatoes is not noticed. However, in the far north, e.g. Finland, the long days of summer can delay tuber initiation. In North Africa the effect of high temperatures may interact with the shorter days to change the characteristics of a 'late' cultivar to be more akin to those of earlier maincrop cultivars.

This provides a nice example of where the question of necessity and sufficiency should be considered by both the developer and the intended user of a DSS. There is not a single clear answer. The 'correct' answer depends entirely on the purposes for which the DSS is to be used.

So, is it necessary to include the effects of photoperiod in a model of potato growth or in a DSS? The correct answer appears to be equivocal: Yes and no! In a DSS such as Agro-Zoning that, it is planned, should be applied to conditions worldwide it is indeed necessary to include them. Yet they are not included in the other models or DSS described in this book, nor in several other extant models of potato growth. Why? It is because most of the other models have been developed and applied over a limited latitudinal range. For such models and any DSS that use them, any algorithms that provide estimates of timing of tuber initiation or of numbers of tubers formed are attuned, however unwittingly, to the conditions of the locality where they were developed and are being used. Where this is recognised and where a model needs to be 'transportable' i.e. to be applied in another area, those algorithms can be modified to reflect the new application. This may be among the effects that are accommodated by the common practice (although not universal) of 'calibrating' a model before applying it. Under such circumstances, it is not necessary to include photoperiod in the models and, indeed, to do so risks introducing unnecessary errors.

Most farmers grow their potatoes in a fairly tightly defined season. In a cool temperate climate, they plant potatoes as soon as soil temperature rises above about 7°C, although difficulties with wet ground conditions can spread planting over a period of six to eight weeks, year-to-year. In such circumstances, the differences in day length after emergence of the crop do not appear to change tuber initiation. - Or, at least, they have not yet been blamed for difficulties with the crop.

On the other hand, consider that differences in the chosen season for growth impose differences in day length at the times when the potato crop might be ready to initiate tubers. For example, planting in Greece is phased across regions and altitudes from early December to mid-July (MacKerron, 1992). The differences in associated day lengths at, say, three weeks after emergence are considerable. In warmer climates, the constraint on the time of planting may be the wait for soil temperatures to fall and in some areas, e.g. in parts of South Africa and in Israel, there may be both autumn-planted and spring-planted crops. In such cases one crop is planted to meet falling temperatures and shortening days and the other to meet rising temperatures and lengthening days. These instances all present cases where is possible or even probable that the effects of photoperiod and its interaction with temperature should be included in a model of potato growth that is to be used within one geographic region.

Should the effects of photoperiod be included in a model of potato growth? - It depends!

Weather variables?

Many DSS for agricultural production are driven by weather variables. The choice of which variables to use might seem obvious but should reflect not only the science but, where there are alternatives, it should also reflect the ease of obtaining the data.

The SCRI model of potato growth and development that underlies part of **MAPP** (Chapter 8 - MacKerron et al.) was planned to run on generally available weather data and so, twenty years ago, these included wet and dry bulb temperatures (T_w, T_a) to quantify humidity. These are two standard weather variables measured in every Stevenson screen in the UK. Now, however, the ideal source of weather data is an automatic weather station (AWS) and these do not measure T_w. Instead they generally use a capacitance sensor that is calibrated on relative humidity. The inputs to the model will have to be changed.

Also, MAPP uses soil temperature at 10 cm depth (T_{10}). This is a variable that is measured in Agro-climatic weather stations but not in the more general climatological weather stations. Is it necessary? It has two purposes. The first is in the estimation of sprout growth to emergence. For estimating time to emergence one could forego T_{10} and use T_a with a derivation relating soil temperature to air temperature - but the parameters of the relation would differ with soil type and soil moisture content. Is it sufficient? The second use of T_{10} is to estimate the rate of root extension. Yet very shortly after the roots are formed, the growing points will be deeper than 10 cm. Then scientific accuracy demands that a whole profile of soil temperatures should be provided at 20, 30, 50, and 100 cm. The judgement of those developing the models underlying MAPP was that, at the present time, such an array of temperature data is too much to ask as a routine. Some kind of accuracy gave way to practicality.

Number of tubers

An influence diagram is very useful to set out those factors and variables that influence the states of development and the rates of processes. Should a model or a DSS include all the factors known to influence a state variable or only those for which the influence has been quantified? Should it include all of them? The influence diagram in Chapter 8, Figure 3 shows several influences on number of tubers that are not treated within MAPP and at least one of these, the effect of soil moisture deficit at around the time of tuber initiation has been quantified (MacKerron & Jefferies, 1986) yet that effect is not simulated within MAPP.

The number of daughter tubers is very important in determining the distribution of tuber sizes and so the value of a crop at any given yield. MAPP includes the effects of cultivar, seed size and seed spacing in providing an estimate of number of tubers that is used to calculate the consequences of a range of seed rates. These are influences that have major effects on the population of daughter tubers and they are almost entirely within the control of the grower. But MAPP only provides an estimate. After planting a number of factors such as photoperiod and size of canopy at tuber initiation and, perhaps, soil strength as well as soil moisture deficit can all influence the number of daughter tubers produced. We cannot know the values of the input factors until after they have happened and then we can see the effects. And, crucially, from the time when tubers are initiated, there is no management task that can modify that number. There is a real crop. It can be examined to see what nature has provided. There is no need for a model at that point. Therefore, the developers of MAPP judged that including a function for the influence of soil water on number of tubers was not worth doing and that it would be simply an example of unnecessary (and unrealistic) accuracy.

Another influence on number of tubers and that could be simulated in MAPP is depth of planting. Depth of planting has a non-linear effect on the number of sub-surface nodes on each stem which, in turn, influences the number of main stolons that are formed. But nodes are integer quantities so a stem has two, then three, then four. If depth of planting were to be coupled in the model through number of nodes to number of stolons and thence to number of tubers it would at best make the model unnecessarily complicated, at worst it could possibly make the model very wrong.

Root growth

In Potato-zoning, roots are expected to grow to 60 cm. That is adequate for the purposes of that DSS.

Both the Potato calculator (Chapter 6 - Jamieson et al.) and MAPP (Chapter 8 - MacKerron et al.) use an index of water stress to modify the estimate of canopy growth and both derive that index from the ratio of the rate of supply of water in the root zone to the evaporative demand for water. Both calculate root extension as a function of thermal time. In the Potato calculator, maximum root depth is set as a parameter of the cultivar. That refinement was avoided in MAPP but an unintended feature was found once MAPP was operational.

In MAPP, the emerging sprout forms nodes as a function of thermal time and the thermal time that is required to elapse before emergence is dependent upon planting depth. So, the number of sub-surface nodes is seen to be influenced by planting depth. - That seems reasonable. And yet, a certain number of roots arise at each sub-surface node. So, a small change in the declared depth of planting can give a switch between three and four sub-surface nodes and a step-change in the number of roots that are formed. This has consequences later for the calculated availability of water and for canopy expansion, and later still, for the development of water stress. There was a case in which a grower entered 8 cm as planting depth and this led to a faulty simulation of crop development

This is an example where a necessary detail (planting depth for its influence on time from planting to emergence) can have unexpected consequences and where 'accuracy' or attempted 'realism' can cause real difficulties.

Other particular examples of features of DSS

Potato-zoning (Chapter 2 - Haverkort et al.) uses an FAO soil map with a pixel grid of 5 x 5 minutes of arc that distinguishes only three classes of soil texture to provide data including availability of soil water and it starts by assuming a soil water content of 80% of field capacity on the day of planting. Can these definitions of soil be adequate in a DSS intended to calculate the potential for production of potatoes? Before answering that question on must appreciate that there is likely to be a range of soil types within the area represented by even one pixel. Further, the climate data that substitutes for weather data is long-term average data based on a pixel grid of 30 x 30 minutes of arc. Certainly, the definitions of the inputs are simple but they are clear and since they refer to large units of land, it can be argued that there is little point in seeking more precise inputs unless they are provided in a finer grid. Potato-zoning, then, produces answers using data that is available. The potential users will judge whether it is sufficient.

If the model of **PCN development** described in Chapter 9 (Elliott et al.) is to be tailored to an individual field and crop it needs to have a sophisticated adjustment of its parameters. This is not a job for a grower but rather one for an advisor. The model in the stage of development in which it has been presented is, therefore, more suited to educate growers on their strategy against PCN rather than to present specific tactics. That is a useful facility but it illustrates the need to consider during development who is the foreseen user and what is the use that he or she will make of the model or DSS.

The **Potato calculator** (Chapter 6 - Jamieson et al.) promises to be a successful DSS on the application of water and nitrogen to the potato crop and intended users have been involved in early testing of its performance. However, it seems to have no provision for monitoring the crop with a view to modifying planned supplementary application of nitrogen.

The particular application of the model **Azobil** by Goffart et al. (Chapter 5) gives a good example of restricted accuracy in the cause of practicality. Nitrogen recovery is assumed to be 100%; but it isn't. The model is additive, without interactions; but there are some. Gaseous N-losses are compensated; that's a coincidence. And there is no leaching under a cropped soil. That is probably true in most cases for most of the growing season. By coupling the fractional application of what is the calculated requirement for nitrogen with a simple set of field measurements, it allows the grower to save on N-fertilizer and reduce the risk of excessive applications that may be leached without reducing yields.

Conclusions

The earlier part of this chapter aimed to remind the reader of the nature of the derivation of the relations underlying most or all of the models used in agricultural science. If a deterministic model can be used to predict with negligible error, for practical purposes, then it is reasonable to use it. If not then we should really seek a probabilistic model, which still would not be an exact characterisation of nature but which would allow us to assess the validity of the inferences that we make. Until such probabilistic models are developed it will be important that DSS based on deterministic relations include adequate crop monitoring to allow the re-scaling of the model's calculations, and that they also show the calculated effects of decisions or inputs that differ incrementally from the value that is given as the 'best' value of the 'explanatory' or independent variable.

MacKerron & Waister (1985) offered a number of applications for the SCRI model of potential yield of potatoes. Models should always have an application. This is the case here with DSS which by their nature, are attempts to apply models to real problems. Other criteria for models should be simplicity of input for 'real-world' applications and adequacy of output for relevance to those applications.

Simplicity of use is highly important and may even be more important than accuracy. A contributory reason for the failure of DSS in the past was the use of command-line models. The advent of graphical user interface (GUI) modelling has changed that as most potential users are well-acquainted with the standard Windows®-type of interface but still a DSS must be usable with almost no training and have readily-acquired inputs or it will not be accepted by the intended users.

Where data has multiple uses, the answer to what is necessary will hinge on the more demanding application. Weather data is a special example of this point. There are models of plant growth that are driven by the calculation of photosynthesis using hourly values of weather, yet the precision obtained from models using simple daily values may be at least as good. On the other hand, many of the weather variables used in models of growth and development are also used in models of risk of disease and of progress of disease which may, indeed, require data to be acquired more frequently e.g. hourly. In such a case the user of both models will need to acquire the more frequent data but also summarise it for the less demanding model.

The DSS presented in this book all start from the modellers' assumption that a deterministic, semi-mechanistic model can give a correct answer, rather than one answer from a likely range of answers. The two most useful advances in decision support would be the more general development of Bayesian (probabilistic) models in place of deterministic ones, coupled with an acceptable means of presenting the results to the general user. Until such time as all models are Bayesian and all users can understand these, the current style of DSS offer the best way forward.

References

MacKerron,D.K.L.and Jefferies,R.A. 1986. The influence of early soil moisture stress on tuber numbers in potato. Potato Research, 29, 299-312.

MacKerron D.K.L. and Waister P.D. 1985. A simple model of potato growth and yield. I. Model development and sensitivity analysis. Agricultural and Forest Meteorology 34, 241-251.

MacKerron D.K.L. 1992. Agrometeorological Aspects of Forecasting Yield of Potato within the European Community. Commission of the European Communities, Luxembourg. (ISBN 92-826-3538-4; Catalogue Number: CD-NA-13909-EN-C) 250pp.

Authors

T.H. Been
Wageningen University and Research Centre, Plant Research International
PO Box 16, 6700 AA Wageningen, The Netherlands

H. Boizard
INRA, Unité d'agronomie Laon-Reims-Mons
F-02007 Laon cedex, France

R. van den Boogaard
Wageningen University and Research Centre, Agrotechnology & Food Innovations
PO Box 17, 6700 AA Wageningen, The Netherlands

R. Booij[†]
Wageningen University and Research Centre, Plant Research International
P.O. Box 16, 6700 AA, Wageningen, TheNetherlands

C. Chambenoit
Agro-Transfert
Domaine de Brunehaut, F-80200 Estrées-Mons, France

M.J. Elliot
Scottish Crop Research Institute
Dundee, DD2 5DA, Scotland, United Kingdom

J.P. Goffart
Centre de Recherches Agronomiques de Gembloux, Département Production Végétale
4 rue du Bordia, 5030 Gembloux, Belgium

C. Grashoff
Wageningen university and Research Centre, Plant Research International
P.O. Box 16, 6700 AA Wageningen, The Netherlands

J. Hadders
Dacom PLANT-Service BV
P.O. Box 2243, 7801 CE Emmen, The Netherlands

A.J. Haverkort
Wageningen university and Research Centre, Plant Research International
P.O. Box 16, 6700 AA Wageningen, The Netherlands

H. Hinds
Plantsystems Ltd
97 Hollycroft Road, Emneth, Wisbech PE14 8BB, United Kingdom

P.D. Jamieson
New Zealand Institute for Crop & Food Research Ltd
PB 4704, Christchurch, New Zealand

C. Kempenaar
Wageningen University and Research Centre, Plant Research International
PO Box 16, 6700 AA Wageningen, The Netherlands

F. Laurent
ARVALIS-Institut du végétal, Station expérimentale
F-91720 Boigneville, France

J.M. Machet
INRA, Unité d'agronomie Laon-Reims-Mons
F-02007 Laon cedex, France

D.K.L. MacKerron
Scottish Crop Research Institute
Invergowrie, Dundee, DD2 5DA, United Kingdom

B. Marshall
Scottish Crop Research Institute
Invergowrie, Dundee DD2 5DA, United Kingdom

D. Martin
Plantsystems Ltd
97 Hollycroft Road, Emneth, Wisbech PE14 8BB, United Kingdom

R.J. Martin
New Zealand Institute for Crop & Food Research Ltd
PB 4704, Christchurch, New Zealand

J.W. McNicol
Scottish Crop Research Institute
Invergowrie, Dundee, DD2 5DA, United Kingdom

M. Olivier
Centre de Recherches Agronomiques de Gembloux, Département Production Végétale
4 rue du Bordia, 5030 Gembloux, Belgium

M.S. Phillips
Scottish Crop Research Institute
Dundee, DD2 5DA, Scotland, United Kingdom

P. Raatjes
Dacom PLANT-Service BV
P.O. Box 2243, 7801 CE Emmen, The Netherlands

C.H. Schomaker
Wageningen University and Research Centre, Plant Research International
PO Box 16, 6700 AA Wageningen, The Netherlands

S. Sinton
New Zealand Institute for Crop & Food Research Ltd
PB 4704, Christchurch, New Zealand

P.J. Stone
New Zealand Institute for Crop & Food Research Ltd
PB 4704, Christchurch, New Zealand

D.L. Trudgill
Scottish Crop Research Institute
Dundee, DD2 5DA, Scotland, United Kingdom

D. Uenk
Wageningen University and Research Centre, Plant Research International
P.O. Box 16, 6700 AA, Wageningen, The Netherlands

P.W.J. Uithol
Wageningen University and Research Centre, Plant Research International
P.O. Box 16, 6700 AA Wageningen, the Netherlands

A. Verhagen
Wageningen University and Research Centre, Plant Research International
P.O. Box 16 6700 AA Wageningen, The Netherlands

M.W Young
Scottish Crop Research Institute
Invergowrie, Dundee DD2 5DA, United Kingdom

R.F. Zyskowski
New Zealand Institute for Crop & Food Research Ltd
PB 4704, Christchurch, New Zealand

Index

A
absorbance – 190
absorption – 187
accuracy – 147, 213, 214
active ingredient – 171
agribusiness – 170
agro-chemical – 23, 24
agro-ecological zoning – 33, 41
agro-ecosystem – 33
Alternaria solani – 170, 171, 181
Andean ecoregion – 36
annual decline rate – 147, 150
annual weeds – 195
aphid – 21, 42, 183
AZOBIL – 54, 55, 69, 71

B
bacteria – 42, 157
baker – 101
balance-sheet method – 69
Bayesian models – 223
bed system – 124
biomass – 56, 86, 91
    accumulation – 85, 86
    partitioning – 85
black spot bruising – 43
breeder – 21
breeding – 199
burn-down – 122
buyer – 21

C
calendar – 130
calibration – 104, 115
canopy – 85, 86, 137
    closure – 131, 210
    decline – 132
    development – 122
    expansion – 86, 87, 131
    senescence – 86
    size – 85

Cara – 109, 177
carbon dioxide – 201
catch crop – 19, 71
centre pivot irrigation – 179
cereals – 200
certification – 22, 24, 155, 156
Charlotte – 107, 108
charts – 147, 148, 150
chemical
    contact – 169
    efficacy – 180
    systemic – 169
    translaminar – 169
    treatment – 142
    wear off – 171
Chenopodium album – 189
chitted – 124
chlorophyll-meter – 68-70, 73, 74, 76-78, 80, 82
chloroplasts – 189
choice of cultivar – 140
climate – 28, 30
climatic hazard – 29, 199
cloud free images – 183
cloudiness – 33
Colorado beetle – 42
commercial rotation – 144
common scab – 131, 135, 137, 176
computer – 17
    interface – 120
    personal – 156
    screen – 65
    software – 101
consultancy firms – 24
consumer – 20, 21
control measure – 154, 161
control strategy analysis – 147
controlling variables – 120
cost/benefit – 161
    analyse – 154
    ratio – 154

Q

Q-organism – 154
quality – 29, 70, 167
    assurance – 22
    assurance system – 155
    control – 22
    requirements – 54
quantitative risk management – 154

R

radiation – 84
rain – 171, 176, 195
rainfall – 38, 89
rainfed condition – 37
recommended dose rate – 171
recovery coefficient – 72
reducing sugars – 43
refill point – 169
reflection – 49, 187
registration – 25, 158
regression analysis – 214
relative humidity – 171, 176
relative leaf expansion rate – 205
relative soil water volume – 176
remote sensing – 22, 183
residual biomass – 88
residual carbon – 200
residual mineral nitrogen – 51
residues – 22
resistance – 142, 145, 146, 154, 164
resource use efficiency – 42
respiration – 210
retailer – 21
retrospective sampling – 151
rice – 200
riddle
    size – 101, 110, 111, 115
    square – 103
risk
    analysis – 25
    disease – 222
    leaching – 97
    perception – 154

root – 32, 61
    death – 132
    depth – 84, 88, 198, 202, 206, 210
    growth – 132, 220
    zone – 85
rooting – 60, 198
rotation – 18, 73, 74, 142, 143, 154, 202
rotational gaps – 145
row spacing – 100, 124
Russet Burbank – 87, 88, 96

S

safety of food products – 20, 155
sampling – 157, 158
screenshot – 139, 161-164, 193, 194, 204, 207
seed
    cost – 129, 130
    price – 126
    rate – 118, 119, 122, 127, 128, 130, 133, 138, 217
    size – 118, 119, 122, 126, 130
    spacing – 130
seed potato – 20, 28
    grower – 21
    trade – 21
seed-rate model – 121
self adjusting system – 25
self learning system – 22, 24
semi-mechanistic model – 120, 123, 215
Senecio vulgaris – 192
senescence – 132, 160, 205
sensing techniques – 23, 186, 187
set points – 50
setting, default – 147
sexual reproduction – 171
Shepody – 108, 110
simulation model – 33, 93
Sirius Wheat Calculator – 86

wholesaler – 21
wilting point – 34, 89
wind – 171
wind speed – 175
winter-grown potatoes – 90

Y
yield
      anticipated – 126
      attainable – 204
      expected – 145
      graded – 121, 122, 128, 131, 133
      index – 57
      potential – 57, 84
      reducing factors – 201
      simulated – 91
      total – 101
yield loss, proportional – 145
yield-gap analysis – 25

Z
zero N-window – 76
zero-window – 75
zoning – 28, 29, 32

Printed in the United States
by Baker & Taylor Publisher Services